Jule Peters

Revision of the genus Fosterella L.B. Sm.

Jule Peters

Revision of the genus Fosterella L.B. Sm.

Bromeliaceae

Südwestdeutscher Verlag für Hochschulschriften

Impressum/Imprint (nur für Deutschland/ only for Germany)
Bibliografische Information der Deutschen Nationalbibliothek: Die Deutsche Nationalbibliothek verzeichnet diese Publikation in der Deutschen Nationalbibliografie; detaillierte bibliografische Daten sind im Internet über http://dnb.d-nb.de abrufbar.

Alle in diesem Buch genannten Marken und Produktnamen unterliegen warenzeichen-, marken- oder patentrechtlichem Schutz bzw. sind Warenzeichen oder eingetragene Warenzeichen der jeweiligen Inhaber. Die Wiedergabe von Marken, Produktnamen, Gebrauchsnamen, Handelsnamen, Warenbezeichnungen u.s.w. in diesem Werk berechtigt auch ohne besondere Kennzeichnung nicht zu der Annahme, dass solche Namen im Sinne der Warenzeichen- und Markenschutzgesetzgebung als frei zu betrachten wären und daher von jedermann benutzt werden dürften.

Verlag: Südwestdeutscher Verlag für Hochschulschriften Aktiengesellschaft & Co. KG
Dudweiler Landstr. 99, 66123 Saarbrücken, Deutschland
Telefon +49 681 37 20 271-1, Telefax +49 681 37 20 271-0
Email: info@svh-verlag.de
Zugl.: Kassel, Universität Kassel, Dissertation, 2009

Herstellung in Deutschland:
Schaltungsdienst Lange o.H.G., Berlin
Books on Demand GmbH, Norderstedt
Reha GmbH, Saarbrücken
Amazon Distribution GmbH, Leipzig
ISBN: 978-3-8381-1456-9

Imprint (only for USA, GB)
Bibliographic information published by the Deutsche Nationalbibliothek: The Deutsche Nationalbibliothek lists this publication in the Deutsche Nationalbibliografie; detailed bibliographic data are available in the Internet at http://dnb.d-nb.de.

Any brand names and product names mentioned in this book are subject to trademark, brand or patent protection and are trademarks or registered trademarks of their respective holders. The use of brand names, product names, common names, trade names, product descriptions etc. even without a particular marking in this works is in no way to be construed to mean that such names may be regarded as unrestricted in respect of trademark and brand protection legislation and could thus be used by anyone.

Publisher: Südwestdeutscher Verlag für Hochschulschriften Aktiengesellschaft & Co. KG
Dudweiler Landstr. 99, 66123 Saarbrücken, Germany
Phone +49 681 37 20 271-1, Fax +49 681 37 20 271-0
Email: info@svh-verlag.de

Printed in the U.S.A.
Printed in the U.K. by (see last page)
ISBN: 978-3-8381-1456-9

Copyright © 2010 by the author and Südwestdeutscher Verlag für Hochschulschriften Aktiengesellschaft & Co. KG and licensors
All rights reserved. Saarbrücken 2010

to my mother

Vom Fachbereich Naturwissenschaften der Universität Kassel im Juli 2009 als Dissertation angenommen.

Dekan: Prof. Dr. Friedrich W. Herberg
Gutachter: Prof. Dr. Pierre L. Ibisch und Prof. Dr. Kurt Weising

Die vorliegende Arbeit wurde unter der Betreuung von Prof. Dr. Pierre L. Ibisch in Kooperation mit der Fachhochschule Eberswalde, Fachbereich für Wald und Umwelt angefertigt.

Acknowledgements

I acknowledge very much the financial support by the University of Kassel. The achievement of this work would not have been possible for me without the provided PhD research fellowship. The field trip to South America was kindly supported by the DAAD.

For the provision of the topic I would like to thank my supervisor Pierre L. Ibisch wholeheartedly. I benefited tremendously from his kind support and friendly advice just like his enthusiasm for *Fosterella* and Bolivia has infected me.

My sincere thanks to my supervisor Kurt Weising for support and confidence throughout the whole project.

I want to thank all further colleagues involved in the *Fosterella* project for the reliable cooperation: Georg Zizka and Katharina Schulte (Research Institute Senckenberg, Frankfurt/M.), Christoph Nowicki (University of Applied Sciences, Eberswalde), Martina Rex and Natascha Wagner (University of Kassel). Special thanks go to Christoph Nowicki for help with the distribution maps.

We are very grateful to our longtime cooperation partners abroad, without whom the whole *Fosterella* project could not have been hardly that successful: Roberto Vásquez (Sociedad Boliviana de Botánica, Santa Cruz, Bolivia), Stephan G. Beck (Herbario Nacional de Bolivia, Instituto de Ecología, La Paz, Bolivia) and Elton M.C. Leme (Herbarium Bradeanum, Rio de Janeiro, Brazil). Thank you so much for your generous help.

I am deeply indebted to several people who made the field trip to South America not only possible and successful, but the most marvelous and impressive expedition of my life: Pierre L. Ibisch, Roberto Vásquez, Raul Lara, Stephan Beck, Alfredo Tupayachi, Javier Farfan-Flores, Jean-Paul Latorre, Betty Millan, Familia Huertas, Alfredo Grau, Figuera Romero, Roberto

Neumann, Lázaro Novara and Elton M.C. Leme. Very special thanks to my companion Nicole Schütz.

Many thanks to all curators and staff members of herbaria and botanical gardens who provided material and related information. Special thanks to Timm Stolten (Botanical Gardens Heidelberg), Jürgen Lautner (Old Botanical Garden Göttingen), Beat Leuenberger and Robert Vogt (Botanic Garden and Botanical Museum Berlin), Nikolai Friesen (Botanical Garden Osnabrück), Eric Gouda (Botanic Gardens Utrecht), Walter Till (Herbarium of the University Vienna), Bruce Holst, Harry Luther and Rosalind Rowe (Marie Selby Botanical Gardens). Special thanks go to Sabine Bringmann for cultivation of the *Fosterella* living collection in Kassel.

I greatly appreciate the kind help in all taxonomical issues provided at any time by Helmut Freitag.

Many thanks for revising the manuscript to Helmut Freitag, Kerstin Volkenant and Anita Fischer.

Warm thanks also to all members of the department of Botany at the University of Kassel, namely Kurt Weising, Helmut Freitag, Frau Maier-Stolte, Irene Diebel, Christine Frohmuth, Tim Kröger, Daniela Guicking and Natascha Wagner. It has been a pleasure for me to join this friendly team. Special thanks again to Nicole Schütz.

There were some more very nice people who made the AVZ a lovely place to work: Horst Koenies, Clovis Douanla-Meli, Kerstin Volkenant, Tina Schäfer, Astrid Schröder, Kerttu Valtanen, Manuel Punzet, Tine Nowack, Toni Burmeister, Daniel Großarth, Matze & Franky, Sabine Bringmann and Ralf Linzert – thank you all.

I am grateful to the cooperative and hospitable people I met during my pleasant stay in Eberswalde: Harald Schill, Bernhard Götz, Ute Krakau, Stefan Kreft, Marcela Cuadros, Camilla, Kolja, Tido, Steffi, Dirk, Sebastian and Franka.

Finally, love to my dear family and friends – Papa, Flori, Lisa, Bastian, Joanna, Holger, Nici, Frauke, Kerstin, Peter, Seb, Aisha, Karel, Heike, Hendrik, Tim, Christin, Tine, Freddi, Jantje, Mia, Antje, Olli, Eri, Luddi, Jan, Sinja, Christian, Linda, Jörn, Sebi, Gerd, Horst, Aki, Kerttu, Tina, Clovis, Anita, Sean, Rolf, Barbara, Jörg, Christian, Ralf, Gaby, Volker – I am glad you are around!

Contents

1 **Introduction** 1
   1.1 The family Bromeliaceae Juss. 1
   1.2 The genus *Fosterella* L.B. Sm. 4
   1.3 Objectives of the study 7

2 **Material and Methods** 9
   2.1 Plant material 9
      a) Herbarium material 9
      b) Living collections 9
      c) Field observations 10
   2.2 Assessment of morphological characters 10
   2.3 Evaluation of biogeographical data 17
   2.4 Reconstruction of character evolution 17

3 **Results** 19
   3.1 *Fosterella* L.B. Sm. 19
   3.2 Key to the species 23
   3.3 Taxonomic treatment 27
      F. albicans (Griseb.) L.B. Sm. 27
      F. aletroides (L.B. Sm.) L.B. Sm. 35
      F. batistana Ibisch, Leme & J. Peters 39
      F. caulescens Rauh 43
      F. chaparensis Ibisch, R. Vásquez & E. Gross 48
      F. christophii Ibisch, R. Vásquez & J. Peters 52
      F. cotacajensis M. Kessler, Ibisch & E. Gross 56
      F. elviragrossiae Ibisch, R. Vásquez & J. Peters 61
      F. floridensis Ibisch, R. Vásquez & E. Gross 65
      F. gracilis (Rusby) L.B. Sm. 70
      F. graminea (L.B. Sm.) L.B. Sm. 74
      F. hatschbachii L.B. Sm. & Read 78
      F. heterophylla Rauh 82
      F. kroemeri Ibisch, R. Vásquez & J. Peters 86

        F. micrantha (Lindl.) L.B. Sm. 90

        F. nicoliana J. Peters & Ibisch 97

        F. pearcei (Baker) L.B. Sm. 101

        F. penduliflora (C.H. Wright) L.B. Sm. 105

        F. petiolata (Mez) L.B. Sm. 114

        F. rexiae Ibisch, R. Vásquez & E. Gross 119

        F. robertreadii Ibisch & J. Peters 123

        F. rojasii (L.B. Sm.) L.B. Sm. 128

        F. rusbyi (Mez) L.B. Sm. 132

        F. schidosperma (Baker) L.B. Sm. 139

        F. spectabilis H. Luther 144

        F. vasquezii E. Gross & Ibisch 148

        F. villosula (Harms) L.B. Sm. 152

        F. weberbaueri (Mez) L.B. Sm. 157

        F. weddelliana (Brongn. ex Baker) L.B. Sm. 162

        F. windischii L.B. Sm. & Read 168

        F. yuvinkae Ibisch, R. Vásquez, E. Gross & S. Reichle 172

   3.4   Species ranges and diversity 178

   3.5   Character evolution 179

4   Discussion 181

   4.1   Collection efforts and taxonomical knowledge 181

   4.2   Range sizes, ecoregions and ecology 182

   4.3   Cytogenetics 182

   4.4   Character evolution 183

   4.5   Conclusions on the evolution and spread of the genus 186

5   Summary 189

6   References 193

Tables

Tab. 1	*Fosterella* specimens subjected to flow cytometry	17
Tab. 2	Reported chromosome numbers in *Fosterella* species	20
Tab. 3	Encoded morphological character states within *Fosterella*	179

Figures

Fig. 1	Molecular phylogeny of Bromeliaceae based on *ndh*F data	3
Fig. 2	Distribution of the genus *Fosterella*	4
Fig. 3	Comparison of the infrageneric phylogenies of *Fosterella* deduced from AFLP analysis and from DNA sequence data	6
Fig. 4	Exemplification of indumentum-types of abaxial leaf surfaces within the genus *Fosterella*	12
Fig. 5	Exemplification of petal shape types within the genus *Fosterella*	16
Fig. 6	Drawing of *Fosterella albicans*	32
Fig. 7	Lectotype of *Fosterella albicans*	33
Fig. 8	*Fosterella albicans* in the natural habitat	34
Fig. 9	Flowers of *Fosterella albicans*	34
Fig. 10	Distribution of *Fosterella albicans*	34
Fig. 11	Drawing of *Fosterella aletroides*	36
Fig. 12	Holotype of *Fosterella aletroides*	37
Fig. 13	Distribution of *Fosterella aletroides*	38
Fig. 14	Drawing of *Fosterella batistana*	40
Fig. 15	Holotype of *Fosterella batistana*	41
Fig. 16	Habit of *Fosterella batistana*	42
Fig. 17	Flowers of *Fosterella batistana*	42
Fig. 18	Distribution of *Fosterella batistana*	42
Fig. 19	Drawing of *Fosterella caulescens*	45
Fig. 20	Holotype of *Fosterella caulescens*	46
Fig. 21	Habit of *Fosterella caulescens*	47
Fig. 22	Flowers of *Fosterella caulescens*	47
Fig. 23	Distribution of *Fosterella caulescens*	47
Fig. 24	Drawing of *Fosterella chaparensis*	49
Fig. 25	Holotype of *Fosterella chaparensis*	50
Fig. 26	Rosette of *Fosterella chaparensis*	51
Fig. 27	Flowers of *Fosterella chaparensis*	51

Fig. 28	Distribution of *Fosterella chaparensis*	51
Fig. 29	Holotype of *Fosterella christophii*	54
Fig. 30	Rosette of *Fosterella christophii*	55
Fig. 31	Flowers of *Fosterella christophii*	55
Fig. 32	Distribution of *Fosterella christophii*	55
Fig. 33	Drawing of *Fosterella cotacajensis*	58
Fig. 34	Holotype of *Fosterella cotacajensis*	59
Fig. 35	Habit of *Fosterella cotacajensis*	60
Fig. 36	Flowers of *Fosterella cotacajensis*	60
Fig. 37	Distribution of *Fosterella cotacajensis*	60
Fig. 38	Holotype of *Fosterella elviragrossiae*	63
Fig. 39	Rosette of *Fosterella elviragrossiae*	64
Fig. 40	Flowers of of *Fosterella elviragrossiae*	64
Fig. 41	Distribution of *Fosterella elviragrossiae*	64
Fig. 42	Drawing of *Fosterella floridensis*	67
Fig. 43	Holotype of *Fosterella floridensis*	68
Fig. 44	*Fosterella floridensis* in the natural habitat	69
Fig. 45	Flowers of *Fosterella floridensis*	69
Fig. 46	Distribution of *Fosterella floridensis*	69
Fig. 47	Lectotype of *Fosterella gracilis*	72
Fig. 48	*Fosterella gracilis* in the natural habitat	73
Fig. 49	Flowers of *Fosterella gracilis*	73
Fig. 50	Distribution of *Fosterella gracilis*	73
Fig. 51	Drawing of *Fosterella graminea*	75
Fig. 52	Holotype of *Fosterella graminea*	76
Fig. 53	Distribution *Fosterella graminea*	77
Fig. 54	Holotype of *Fosterella hatschbachii*	80
Fig. 55	*Fosterella hatschbachii* in the natural habitat	81
Fig. 56	Flowers of *Fosterella hatschbachii*	81
Fig. 57	Distribution of *Fosterella hatschbachii*	81
Fig. 58	Drawing of *Fosterella heterophylla*	83
Fig. 59	Holotype of *Fosterella heterophylla*	84
Fig. 60	Habit of *Fosterella heterophylla*	85
Fig. 61	Flowers of *Fosterella heterophylla*	85
Fig. 62	Distribution of *Fosterella heterophylla*	85
Fig. 63	Drawing of *Fosterella kroemeri*	87

Fig. 64	Holotype of *Fosterella kroemeri*	88
Fig. 65	Rosette of *Fosterella kroemeri*	89
Fig. 66	Flowers of *Fosterella kroemeri*	89
Fig. 67	Distribution of *Fosterella kroemeri*	89
Fig. 68	Drawing of *Fosterella micrantha*	94
Fig. 69	Holotype of *Fosterella micrantha*	95
Fig. 70	Habit of *Fosterella micrantha*	96
Fig. 71	Flowers of *Fosterella micrantha*	96
Fig. 72	Distribution of *Fosterella micrantha*	96
Fig. 73	Drawing of *Fosterella nicoliana*	98
Fig. 74	Holotype of *Fosterella nicoliana*	99
Fig. 75	Distribution of *Fosterella nicoliana*	100
Fig. 76	Holotype of *Fosterella pearcei*	103
Fig. 77	Distribution of *Fosterella pearcei*	104
Fig. 78	Drawing of *Fosterella penduliflora*	111
Fig. 79	Lectotype of *Fosterella penduliflora*	112
Fig. 80	*Fosterella penduliflora* in the natural habitat	113
Fig. 81	Flowers of *Fosterella penduliflora*	113
Fig. 82	Distribution of *Fosterella penduliflora*	113
Fig. 83	Drawing of *Fosterella petiolata*	116
Fig. 84	Holotype of *Fosterella petiolata*	117
Fig. 85	*Fosterella petiolata* in the natural habitat	118
Fig. 86	Flowers of *Fosterella petiolata*	118
Fig. 87	Distribution of *Fosterella petiolata*	118
Fig. 88	Drawing of *Fosterella rexiae*	120
Fig. 89	Holotype of *Fosterella rexiae*	121
Fig. 90	Rosette of *Fosterella rexiae*	122
Fig. 91	Flowers of *Fosterella rexiae*	122
Fig. 92	Distribution of *Fosterella rexiae*	122
Fig. 93	Holotype of *Fosterella robertreadii*	126
Fig. 94	*Fosterella robertreadii* in the natural habitat	127
Fig. 95	Flowers of *Fosterella robertreadii*	127
Fig. 96	Distribution of *Fosterella robertreadii*	127
Fig. 97	Drawing of *Fosterella rojasii*	129
Fig. 98	Holotype of *Fosterella rojasii*	130
Fig. 99	Distribution of *Fosterella rojasii*	131

Fig. 100	Drawing of *Fosterella rusbyi*	136
Fig. 101	Lectotype of *Fosterella rusbyi*	137
Fig. 102	*Fosterella rusbyi* in the natural habitat	138
Fig. 103	Flowers of *Fosterella rusbyi*	138
Fig. 104	Distribution of *Fosterella rusbyi*	138
Fig. 105	Lectotype of *Fosterella schidosperma*	142
Fig. 106	Distribution of *Fosterella schidosperma*	143
Fig. 107	Drawing of *Fosterella spectabilis*	145
Fig. 108	Holotype of *Fosterella spectabilis*	146
Fig. 109	Habit of *Fosterella spectabilis*	147
Fig. 110	Flowers of *Fosterella spectabilis*	147
Fig. 111	Distribution of *Fosterella spectabilis*	147
Fig. 112	Holotype of *Fosterella vasquezii*	150
Fig. 113	*Fosterella vasquezii* in the natural habitat	151
Fig. 114	Rosette of *Fosterella vasquezii*	151
Fig. 115	Distribution of *Fosterella vasquezii*	151
Fig. 116	Lectotype of *Fosterella villosula*	155
Fig. 117	*Fosterella villosula* in the natural habitat	156
Fig. 118	Flowers of *Fosterella villsoula*	156
Fig. 119	Distribution of *Fosterella villosula*	156
Fig. 120	Holotype of *Fosterella weberbaueri*	160
Fig. 121	Habit of *Fosterella weberbaueri*	161
Fig. 122	Flowers *Fosterella weberbaueri*	161
Fig. 123	Distribution of *Fosterella weberbaueri*	161
Fig. 124	Drawing of *Fosterella weddelliana*	165
Fig. 125	Lectotype of *Fosterella weddelliana*	166
Fig. 126	*Fosterella weddelliana* in the natural habitat	167
Fig. 127	Flowers of *Fosterella weddelliana*	167
Fig. 128	Distribution of *Fosterella weddelliana*	167
Fig. 129	Holotype of *Fosterella windischii*	170
Fig. 130	*Fosterella windischii* in the natural habitat	171
Fig. 131	Flowers of *Fosterella windischii*	171
Fig. 132	Distribution of *Fosterella windischii*	171
Fig. 133	Drawing of *Fosterella yuvinkae*	174
Fig. 134	Holotype of *Fosterella yuvinkae*	175
Fig. 135	Habit of *Fosterella yuvinkae*	176

Fig. 136	Flowers of *Fosterella yuvinkae*	176
Fig. 137	Distribution of *Fosterella yuvinkae*	176
Fig. 138	Inference of character evolution in *Fosterella*	180

1 Introduction

1.1 The Family Bromeliaceae Juss.

The neotropical family of Bromeliaceae comprises 58 genera and more than 3100 species (SMITH & TILL 1998, LUTHER 2008). They are distributed in the New World from the Commonwealth of Virginia in the North to Patagonia in the South, one single species occurs in West Africa (*Pitcairnia feliciana*, POREMBSKI & BARTHLOTT 1999).

Bromeliads are perennial, rosette herbs with a short axis or rarely an elongated stem. The roots are absorbing in terrestrial species but in epiphytes their vascular system is reduced and water uptake is accomplished by absorptive trichomes. The leaves are spirally arranged, undivided and entire to spinose; they are usually lepidote by peltate trichomes and frequently forming a water-collecting funnel, a so-called tank. Inflorescences are terminal, sessile or pedunculate, simple to paniculate and often bear conspicuously coloured bracts. The flowers are generally actinomorphic and perfect, rarely somewhat zygomorphic (e.g. *Pitcairnia*) and likewise rarely functionally unisexual (e.g. *Hechtia*). The perianth consists of 3 sepals and 3 petals, each free or connate to each other at the base. The petals frequently are bright coloured and sometimes have basal appendages. The stamens are 3 + 3 in number with the filaments free or adnate to the petals. The anthers are basifixed, introrse and longitudinal dehiscing. The superior or inferior ovary is trilocular and the style slender, with a 3-lobed stigma. The axile placentae bear anatropous ovules which sometimes have chalazal appendages. Septal nectaries are present. The fruit is a septicidal capsule or a berry, the seeds are rather small. They are winged, plumose or naked, containing a cylindrical embryo at the base of a starchy endosperm (DAHLGREN 1985, SMITH & TILL 1998).

Chromosome counts for Bromeliaceae species have been carried out by several authors, revealing $x = 25$ as the basic number for the family (MARCHANT 1967, MCWILLIAMS 1974, BROWN & GILMARTIN 1984 a, 1989 a, VARADARAJAN & BROWN 1985, BROWN et al. 1997, COTIAS DE OLIVEIRA et al. 2004, PALMA-SILVA et al. 2004). Polyploidy is relatively uncommon but has occasionally been reported (LINDSCHAU 1933, DOUTRELIGNE 1939, MARCHANT 1967, DELAY 1974 a, b, BROWN & GILMARTIN 1986, BROWN et al. 1997, COTIAS DE OLIVEIRA et al. 2000, GITAÍ et al. 2005).

Crassulacean acid metabolism (CAM) is very common within the family: MARTIN (1994) estimated two thirds of all bromeliads to be CAM plants, CRAYN et al. (2004) found amongst 1873 investigated taxa of Bromeliaceae 826 CAM species (44%). However, the C3 photosynthetic pathway is inferred to be the ancestral state within the genus, and CAM photosynthesis arose at least four times within Bromeliaceae (CRAYN et al. 2004, GIVNISH et al. 2007).

Bromeliads show a great variety of terrestrial, saxicolous or epiphytic life forms. The family is a prime example for colonisation of extreme habitats like e.g. coastal plains, humid montane forests and high Andean savannahs, and many species are particularly well adopted to xeric conditions (GIVNISH et al. 1997, BENZING 2000, CRAYN et al. 2004). Outstanding diversification in Bromeliaceae is ascribed to several key innovations like the "tank-habit" with water- and nutrient-impounding phytotelmata, absorptive epidermal trichomes, succulence and CAM photosynthesis (BENZING 2000, CRAYN et al. 2004). About half of all Bromeliaceae are epiphytes and apart from the Orchidaceae, they make up the largest portion of epiphytic vascular plants within the Neotropics (BENZING 1990). In arid to semiarid ecoregions like Gran Chaco or Inter-Andean-Dry-Forest, they even constitute 73–82 % of all epiphytic species (IBISCH 1996).

According to recent molecular studies, Bromeliaceae belong to the order Poales within the commelinid clade of the monocots (CHASE et al. 2000, APG 2003, CHASE et al. 2004, GIVNISH et al. 2004). Within the Poales, Bromeliaceae are sister to Typhaceae which in turn are sister to the rest of the order (CHASE et al. 2006, GIVNISH et al. 2006, GRAHAM et al. 2006).

Monophyly of the family has been commonly accepted on the basis of morphological and anatomical characters (DAHLGREN et al. 1985, GILMARTIN & BROWN 1987) and was confirmed by all molecular studies (e.g. CLARK et al. 1993, DUVALL et al. 1993, CRAYN et al. 2004, GIVNISH et al. 2004, 2007). Traditionally, Bromeliaceae have been divided into three subfamilies (Pitcairnioideae, Tillandsioideae and Bromelioideae) that were distinguished by morphological characters of the flowers, fruits and seeds (MEZ 1934, SMITH & DOWNS 1974–1979, SMITH & TILL 1998). However, this view has been challenged by molecular studies (see below).

Quite a few morphological characters have been studied in terms of systematic relevance within the family Bromeliaceae. Among these are, e.g., septal nectaries (BUDNOWSKI 1922, BÖHME 1988), pollen (EHLER & SCHILL 1973, HALBRITTER 1992), stigmata (BROWN & GILMARTIN 1984 b, 1989 b, SCHILL et al. 1988), foliar scales (EHLER 1977, VARADARAJAN & GILMARTIN 1987, PIERCE et al. 2001), seeds (GROSS 1988, 1992, VARADARAJAN & GILMARTIN 1988 a), petal appendages (BROWN & TERRY 1992) and leaf anatomy (HORRES & ZIZKA 1995, PATZOLT 2005). Furthermore, physiological characters (MARTIN 1994, BENZING 2000, CRAYN et al. 2004), chromosome numbers (MARCHANT 1967) and biogeographical data (SMITH 1934 a, 1962) have been analysed regarding their systematic significance. HARMS (1930) established the monotypic subfamily Navioideae, but this did not receive high approval and was not adopted by SMITH & DOWNS (1974–1979). VARADARAJAN & GILMARTIN (1988 b) detected three supposedly monophyletic groups within the subfamily Pitcairnioideae by a cladistic study which was based on foliar anatomy, foliar scale structure, and gross morphology of the shoot and the inflorescence. Accordingly, they proposed tribal rank to the three groups: Brocchinieae, Pitcairnieae and Puyeae (VARADARAJAN & GILMARTIN 1988 c).

Recent molecular studies based on chloroplast DNA sequences revealed that the traditional subfamilies Tillandsioideae and Bromelicideae are monophyletic, whereas Pitcairnioideae are clearly paraphyletic (e.g. RANKER et al. 1990, TERRY et al. 1997, HORRES et al. 2000, 2007, CRAYN et al. 2004, GIVNISH et al. 2004, SCHULTE et al. 2005). Based on a well-resolved *ndh*F phylogeny, GIVNISH et al. (2007) established a new concept of eight bromeliad subfamilies: Tillandsioideae, Bromelioideae, Brocchinioideae, Lindmanioideae, Hechtioideae, Puyoideae, Navioideae and Pitcairnioideae *s.str.* (Fig. 1).

GIVNISH et al. (2007) also made an attempt to date the diversification of the family. The *ndh*F phylogeny was calibrated against the times of origin of other monocot groups based on an *ndh*F phylogeny across monocots, and on the inferred divergence time of *Acorus* from all other monocots 134 million years ago (Mya) (BREMER 2000). According to the molecular clock studies, Pitcairnioideae *s.str.* originated approx. 12.7 Mya, while the radiation of *Fosterella* was estimated to have started 11.5 Mya.

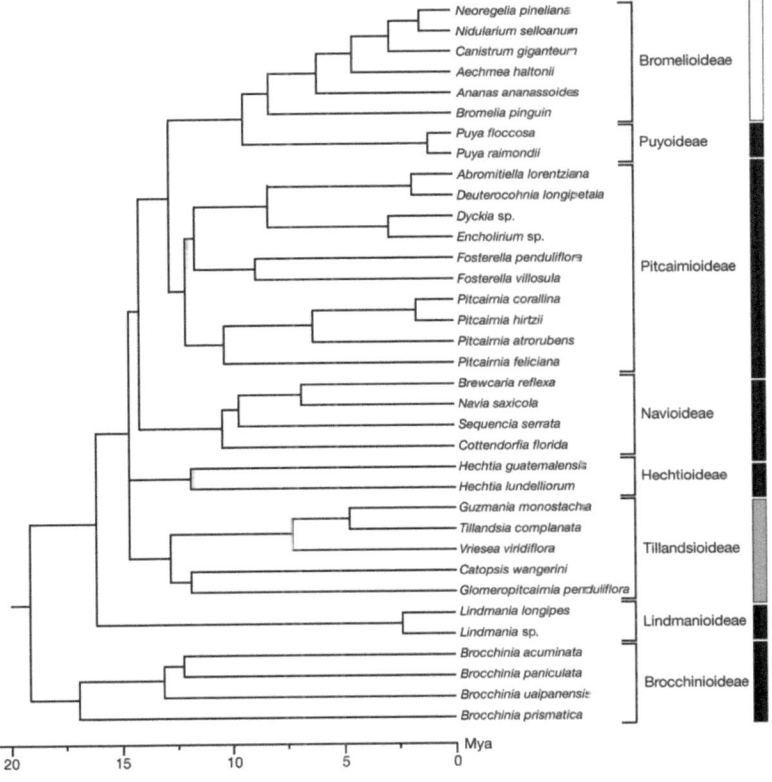

Fig. 1: Molecular phylogeny of Bromeliaceae based on *ndh*F data. Assignment of the new subfamilies to the traditional subfamilies is indicated by shaded bars. Hollow bar: Bromelioideae, grey bar: Tillandsioideae, solid bars: Pitcairnioideae. According to GIVNISH et al. (2007).

1.2 The genus *Fosterella* L.B. Sm.

The genus *Fosterella* traditionally has been placed in the subfamily Pitcairnioideae and remained within Pitcairnioideae *s.str.* according to GIVNISH et al. (2007). *Fosterella* proved to be monophyletic in all molecular studies (e.g. CRAYN et al. 2004, REX et al. 2009). Within Pitcairnioideae *s.str.*, *Fosterella* apparently is sister to a clade comprising the xeromorphic genera *Dyckia*, *Deuterocohnia*, *Encholirium* and finally to the large genus *Pitcairnia* (CRAYN et al. 2004, GIVNISH et al. 2007, REX et al. 2009). *Fosterella* comprises 31 species which are distributed from the Peruvian Andes in the North to northern Argentina and Paraguay in the South. The centre of distribution and diversity is located in arid to semihumid habitats of the northeastern Andean slopes of Bolivia. Two species show a disjunct occurrence: *Fosterella micrantha* in Central America (Mexico, El Salvador & Guatemala) and *F. batistana* in the Brazilian Amazon Basin (Fig. 2).

Fig. 2: Distribution of the genus *Fosterella* based on IBISCH et al. (1999).

Fosterella species are terrestrial, meso- to xerophytic herbs with rosette leaves and pedunculate inflorescences bearing inconspicuous, mostly whitish flowers. Fruits are septicidal capsules that release large numbers of minute, bicaudate seeds. The inner filaments are adnate to the petals, which are not appendaged. The plants are phenotypically rather unspectacular, and are therefore of little interest to amateur collectors and Botanical gardens. Many species are rare, narrow endemics in the sense of MORAWETZ & RAEDIG (2007).

Since the plant architecture of these terrestrial, more or less mesophytic species follows the fundamental monocot pattern with foliage not performing absorptive function, *Fosterella* species belong to the ecological type I according to BENZING (2000). As in other Bromeliaceae, the basic chromosome number is $x = 25$, but polyploidy seems to be comparatively common (LINDSCHAU 1933, DOUTRELIGNE 1939, MARCHANT 1967, DELAY 1974 a, b, BROWN & GILMARTIN 1984 a, 1986, 1989 a, BROWN et al. 1997). All *Fosterella* species investigated so far conduct C3 photosynthesis, whereas all species of its sister genera *Dyckia*, *Deuterocohnia* and *Encholirium*, are CAM plants (MARTIN 1994, CRAYN et al. 2004).

The continuous and interdisciplinary efforts of three working groups (Prof. Weising/Kassel, Prof. Ibisch/Eberswalde and Prof. Zizka/Frankfurt/M.) resulted in the reconstruction of phylogenetic trees based on amplified fragment length polymorphism (AFLP) analysis as well as on chloroplast DNA sequence variation (REX et al. 2007, 2009). These trees provide us with a well-founded molecular framework concerning systematic relationships within *Fosterella*. Within the AFLP study, a neighbour-joining tree based on 77 *Fosterella* specimens, covering 18 species, revealed twelve species groups (A to L) with various levels of bootstrap support. Phylogenetic relationships between these groups remained largely ambiguous though, due to short internal branch lengths. In the phylogenetic analysis based on chloroplast DNA sequence data from the *mat*K gene, *rps*16 intron, *atp*B-*rbc*L and *psb*B-*psb*H intergenic spacers, 96 accessions were included, corresponding to 60 species from 18 genera of Bromeliaceae. Among these, 57 accessions represented 22 *Fosterella* species. Within *Fosterella*, six distinct lineages could be discerned that received high bootstrap support in maximum parsimony trees, and high posterior probabilities in Bayesian analyses. These species groups were refered to as *albicans*, *rusbyi*, *micrantha*, *weberbaueri*, *weddeliana* and *penduliflora* group, respectively (REX et al. 2009).

A comparison of the infrageneric phylogenies deduced from the AFLP analysis and from cpDNA sequence data is shown in Figure 3. Comparing the trees was somewhat difficult because of the uneven taxon representation in both samples. Nonetheless, the topologies were generally congruent, and the six clades defined in the cpDNA tree roughly corresponded to combinations of groups defined by AFLPs. Monophyly of species groups and relationships between groups were generally better resolved by the chloroplast trees, whereas resolution within groups was higher in the AFLP trees. Morphologically defined species boundaries were generally confirmed by molecular analyses.

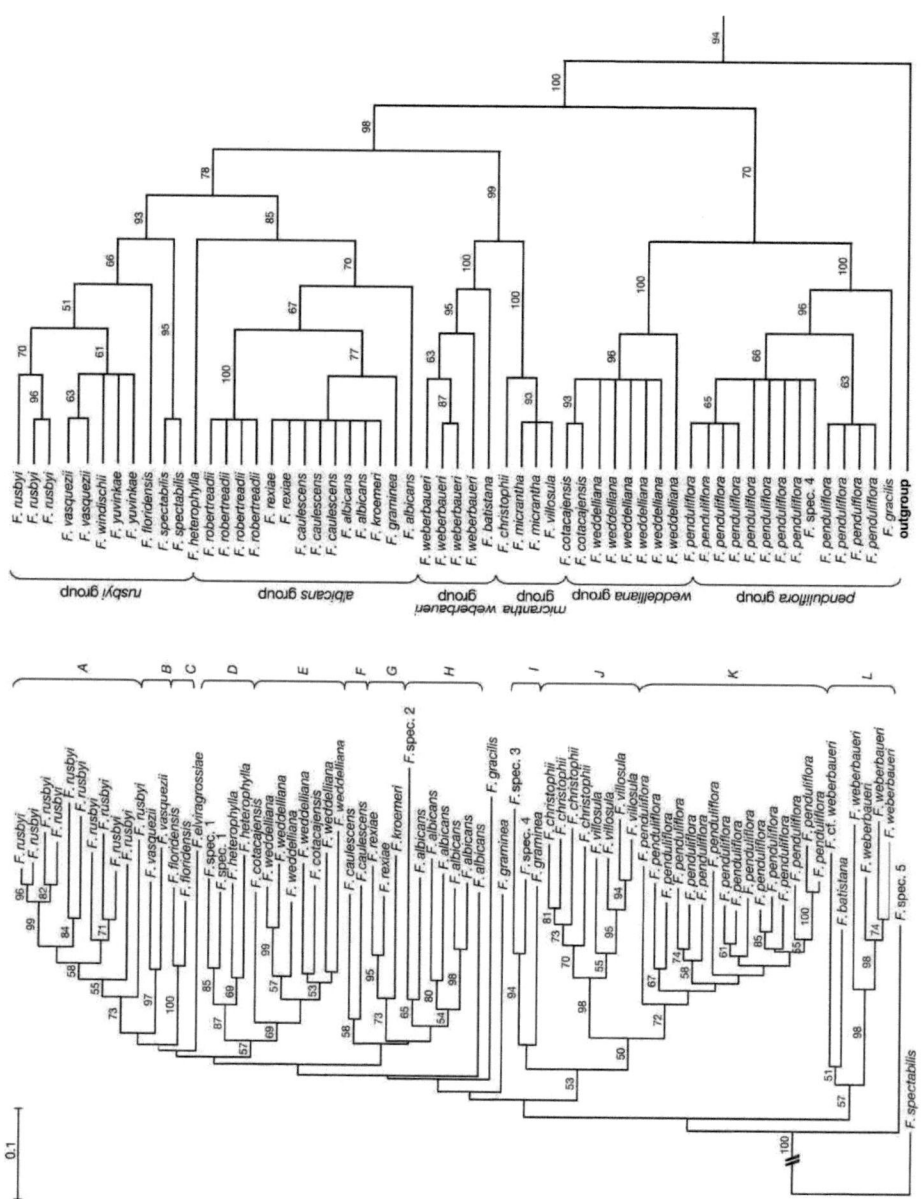

Fig. 3: Comparison of the infrageneric phylogenies of *Fosterella* deduced from AFLP analysis and from DNA sequence data at four chloroplast loci (adopted from REX et al. 2009).

1.3 Objectives of the study

The most recent monograph of the genus *Fosterella* dates back to the work of SMITH & DOWNS (1974), who recognized only 13 species. Since then, numerous additional *Fosterella* taxa have been described (RAUH 1979, 1987, LUTHER 1979, 1997, SMITH & READ 1992, IBISCH et al. 1997, 1999, 2002, KESSLER et al. 1999), raising the number of species known at the beginning of the present work to 30. The most recent key for the identification of the species of *Fosterella* (SMITH & READ 1992) comprises only half of these species. Thus, the intense collection efforts, large number of new taxa, and the considerable progress made in biogeographical and molecular systematic research during the last decades clearly argue for a modern revision of the genus. A well-founded taxonomic concept of *Fosterella* is also of utmost importance for future research, given that the genus provides an excellent model system for investigating speciation mechanisms of small-scale endemics in the Andes.

The intentions of this work are

- To revise the taxonomic concept of the genus *Fosterella* by traditional herbarium methods in comparison with the molecular phylogenies at hand.

- To provide detailed descriptions of all currently recognised *Fosterella* species, including information on etymology, distribution, ecology, taxonomic delimitation and systematic relationships.

- To establish a key for the identification of all currently accepted species of *Fosterella* based on easily detectable characters.

- To reconstruct the evolution of selected morphological characters within *Fosterella*.

- To draw conclusions regarding the evolution and spread of the genus *Fosterella* and its species across their current distributional range.

Some aspects of the present thesis have already become part of scientific publications, including the key to the *Fosterella* species (IBISCH et al. 2008), preliminary aspects of the taxonomical revision (IBISCH et al. 2006, PETERS et al. 2008 b), description of new species (PETERS et al. 2008 a, b) and the mapping of character states onto molecular trees (REX et al. 2009).

2 Material and Methods

2.1 Plant material

The revision of the genus *Fosterella* is based on herbarium material, living collections and observations in the natural habitat, as specified in more detail below. The nomenclature of the genera follows SMITH & TILL (1998), the nomenclature of the subfamilies within the Bromeliaceae follows GIVNISH et al. (2007). Standard forms of authors are used according to BRUMMIT & POWELL (1992). The abbreviations of botanical journals comply with the "Botanico-Perodicum-Huntianum" (LAWRENCE et al. 1968), the abbreviations of monographs with "Taxonomic Literature" (STAFLEU & COWAN 1976–1988). Herbarium acronyms refer to the "Index Herbariorum" (HOLMGREN et al. 1990).

a) Herbarium material

Altogether, more than 800 herbarium specimens, including duplicates, were investigated, most of them in the course of a loan, some of them on-site and a few based on digital high-resolution photographs. The following herbaria kindly provided material on loan: B (15), BA (4), BM (3), F (53), FR (55), GH (28), GOET (6), HB (6), HBG (10), HEID (22), K (16), LI (4), LPB (72), M (6), MO (56), NY (43), SEL (154), U (9), US (28), WU (92).

Some herbaria were visited during a field trip and the respective specimens were investigated on-site: CUZ (5), HSB (9), LIL (16), MCNS (7), RB (4), USM (5), USZ (24).

A few specimens could not be sent on loan – either to conserve type material or to reduce costs – but high-resolution photographs were provided: BM (3), CGE (1), CORD (4), EAP (4), G (2), GZU (1) HB (3), K (5), LAGU (11), MA (3), MEXU (3), MHES (1), P (1), SEL (11), U (1), US (8), WU (13). The following herbaria kindly provided information but do not preserve any *Fosterella* specimens: BIGU, BRH, CR, HUA, INB, MEDEL, USJ, UVAL.

b) Living collections

Besides the herbarium material, living plants represent a most beneficial complement for taxonomic investigations. The following living collections were visited and provided living material – abbreviations as used below in brackets:

Botanic Garden and Botanical Museum Berlin-Dahlem [BGB], Palmengarten Frankfurt/M. [FRP], Old Botanical Garden Göttingen [BGGÖ], Botanical Garden Klein Flottbek, Hamburg [BGHH], Botanical Gardens Heidelberg [BGHD] (all Germany), Botanical Garden Vienna [HBV] (Austria), Fundación Amigos de la Naturaleza [FAN] (Santa Cruz de la Sierra, Bolivia), Refúgio dos Gravatás, private living collection of Des. Elton M.C. Leme [LEME] (Teresópolis, Rio de Janeiro, Brazil).

Leaf material, photographs and related data were kindly provided by the Botanical Garden München-Nymphenburg [BGM], Botanical Garden Osnabrück [BGOS], Botanical Garden Wilhelma Stuttgart [BGS] (all Germany), Botanical Garden Utrecht [BGU] (Netherlands), Royal Botanic Gardens Kew [KEW] (Richmond, Great Britain), Marie Selby Botanical Gardens [MSBG] (Sarasota, Florida, USA). Due to the generous support, a comprehensive living-collection of *Fosterella* plants could be assembled in the greenhouse of the University of Kassel, comprising 128 accessions from 19 *Fosterella* species. This collection will subsequently be transferred to the Botanical Gardens Heidelberg, where an extensive living-collection of Bromeliaceae is located.

c)   Field observations

The studies were further complemented by personal observations in the field, which were made in the course of a three-month field trip to Bolivia, southern Peru, and northern Argentina in autumn 2006. The field trip allowed me to study the morphological variation of *Fosterella* species in their natural habitat. Furthermore, about 150 specimens (including duplicates) were collected, representing nine *Fosterella* species (*F. albicans, F. caulescens, F. penduliflora, F. petiolata, F. robertreadii, F. rusbyi, F. spectabilis, F. villosula, F. weddelliana*).

2.2   Assessment of morphological characters

Morphological observations and measurements were made using a WILD M5A stereomicroscope and an fix-focus magnifier (8x) with scale. For species descriptions and the establishment of a key, characters were selected that are easily detectable and scorable. They were determined by careful inspection of the herbarium vouchers as well as of living plants according to the concept defined below.

Growth form and height

Caulescent means having a conspicuous stem of more than 5 cm in height. This character can only be evaluated in older living plants, because a stem is developed only at an age of three or more years, and stems are in most cases not depicted in herbarium specimens. *Subcaulescent* means having a shorter, rather inconspicuous stem of less than 5 cm in height, and *acaulescent* means not having a stem but a rosette adjacent to the ground. The *height* of the plant is measured when the plant is in flower, from the ground up to the top of the fully developed inflorescence.

Leaves and leaf rosettes

The *number* of leaves is preferably determined in living material, because in most herbarium vouchers only a few leaves are provided. *Few* leaves means up to fifteen leaves per rosette, *many* leaves means more than fifteen. The habit of the rosette is classified as follows: a rosette with the leaves more or less adjacent to the ground is referred to as *flat*. *Arched* rosettes possess arcuated leaves that are not adjacent to the ground over their full length, but only touch the ground with the leaf tips. *More or less upright* rosette means vertically arranged leaves which are not in contact with the ground at all. *Open* rosettes consist of a few, scattered leaves with the ground visible in-between. *Dense* rosettes comprise many leaves in close aggregation, without any space between them.

Leaf sheaths

The *width* of the sheaths is measured at the base, so that the broadest extent is recorded. *Entire* vs. *serrate* refers to the sheath margin, *whitish, greenish, reddish, tinged* or *spotted reddish* refers to the colour and *glabrous* vs. *villous* refers to the abaxial surface of the sheaths.

Leaf blades and indumentum

The following terms describing the shapes of the leaves and leaf apices are used according to STEARN (2004). *Linear:* narrow, with the two opposite margins parallel. *Triangular:* having the form of a triangle. *Lanceolate:* narrowly elliptical, tapering to each end, with the broadest diameter in the basal half. *Oblanceolate:* narrowly elliptical, tapering to each end, with the broadest diameter in the upper half. *Petiolate:* having a petiole. *Acuminate:* terminating gradually in a point. *Acute:* terminating suddenly in a point. *Cuspidate:* tapering gradually into a rigid point. Sizes of the leaf blades generally refer to fully developed leaves of adult plants, the *length* is measured from the base of the blade up to the leaf-tip, the *width* is measured at the broadest sector of each leaf blade. Concerning the thickness, the leaf blades are classified in *thin* (< 3 mm) vs. *succulent* (> 3 mm). The terms *entire, serrate* and *undulate* refer to the margin of the leaf blade.

The classification of the indumentum of the abaxial leaf surface takes into account both macroscopic and microscopic characteristics and is differentiated into four types (Fig. 4):

(a) *Covered by a thick layer of interwoven, peltate trichomes with a fimbriate margin.*
(b) *Densely appressed lepidote by peltate trichomes with a fimbriate margin.*
(c) *Scattered lepidote by peltate trichomes with a dentate margin.*
(d) *Tomentose by stellate trichomes.*

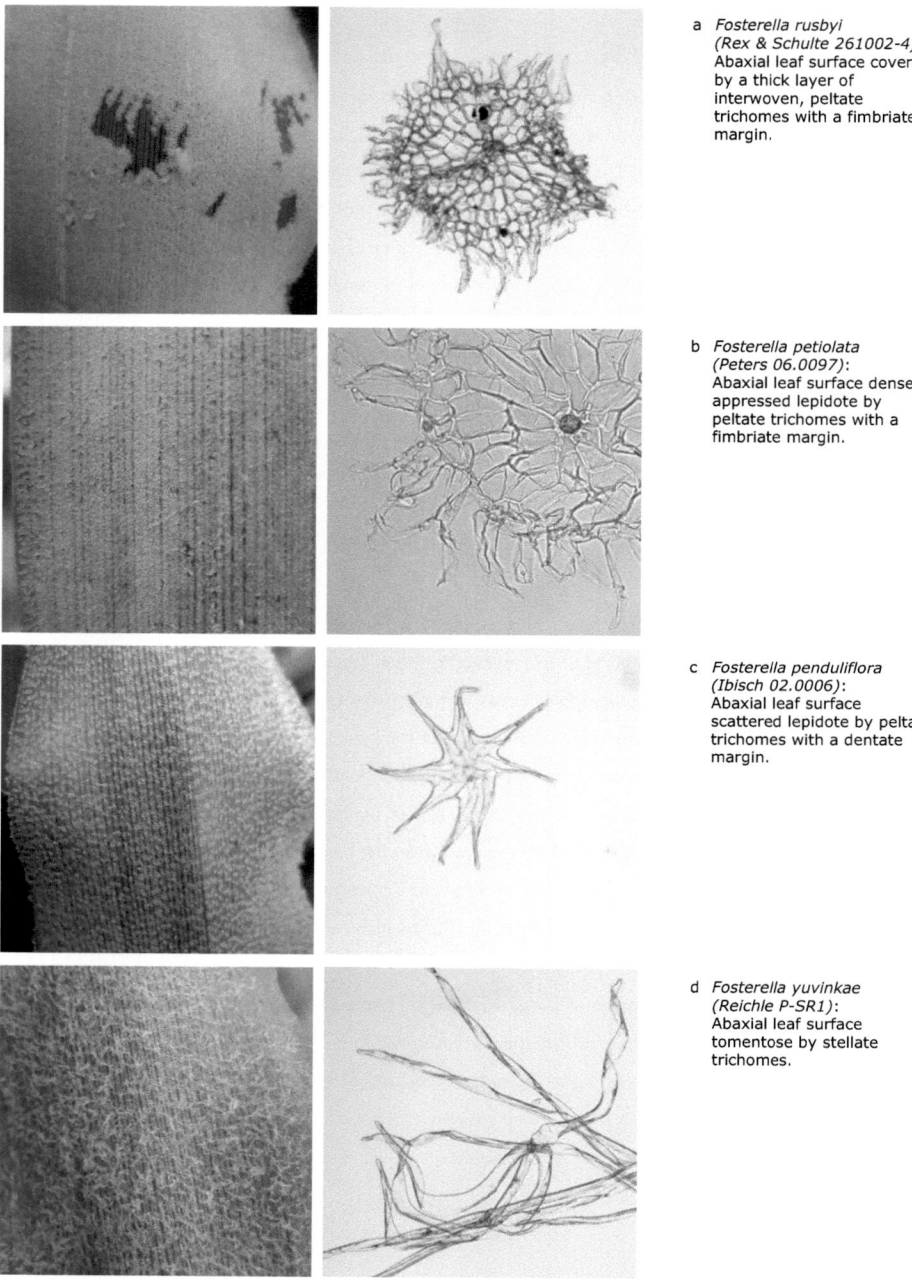

Fig. 4: Exemplification of indumentum types of abaxial leaf surfaces within the genus *Fosterella*. Microscopic photographs of single trichomes by Doreen Karl (KARL 2008).

a *Fosterella rusbyi* (Rex & Schulte 261002-4): Abaxial leaf surface covered by a thick layer of interwoven, peltate trichomes with a fimbriate margin.

b *Fosterella petiolata* (Peters 06.0097): Abaxial leaf surface densely appressed lepidote by peltate trichomes with a fimbriate margin.

c *Fosterella penduliflora* (Ibisch 02.0006): Abaxial leaf surface scattered lepidote by peltate trichomes with a dentate margin.

d *Fosterella yuvinkae* (Reichle P-SR1): Abaxial leaf surface tomentose by stellate trichomes.

Peduncle

The *length* of the peduncle is measured from the base, where it arises from the rosette to the lowest floral bract (in case of racemes) or to the primary bract (in case of compound racemes and panicles). The *diameter* is taken at half height of the scape. The terms describing the indumentum are used following SIMPSON (2006). *Glabrous:* not having trichomes at all. *Glabrescent:* becoming glabrous with time by the loss of trichomes. *Villous:* covered with long, soft, crooked trichomes. *Arachnoid:* having trichomes that form a dense, cobwebby mass. *Glaucous:* covered with a smooth, whitish, waxy coating.

Peduncle bracts

The indicated *length* of the peduncle bracts and internode-to-bract ratio always refers to the ones at half height of the scape. The terms of describing the *indumentum* of the peduncle bracts are the same as for the leaves.

Inflorescence

According to the extent of ramification, the following inflorescence types are differentiated: *racemose* (single axis bearing pedicellate flowers), *compound racemose* (main axis bearing racemes as secondary axes) and *paniculate with branches up to $2^{nd}/3^{rd}$ order* (with several branched axes, each bearing pedicellate flowers). The *length* of the inflorescence is measured from the lowest floral bract (in case of racemes) or primary bract (in case of compound racemes and panicles) up to the top of the fully developed inflorescence. The terms describing the indumentum of the inflorescence are the same as for the peduncle.

Primary bracts

The *length* of the primary bracts and sterile base-to-bract ratio always refers to the ones at half height of the inflorescence. The terms describing the indumentum of the primary bracts are the same as for the leaves.

Branches

The total *number* of branches is given. The terms describing the orientation of the branches are used according to SIMPSON (2006). *Ascending* means directed upward, with a divergence angle of 15°–45° from the vertical axis. *Inclined* means directed upward, with a divergence angle of 15°–45° from the horizontal axis. *Arcuate* means curved, contrary to *straight*. The *length* is taken from the axilla up to the top of the longest branches. The maximum *number* of flowers per branch is given.

Secondary branches

The *length* is taken from the axilla up to the top of the longest secondary branches. The maximum *number* of flowers per secondary branch is given.

Floral bracts

The *length* of the floral bracts and pedicel-to-floral bract ratio always refers to the ones at half length of the branches at half height of the inflorescence. The terms of describing the *indumentum* of the floral bracts are the same as for the leaves.

Flowers

The terms describing the arrangement of the flowers along the inflorescence are used according to STEARN (2004). *Secund:* having flowers turned towards the same side. *Spreading:* sticking out to all directions. *Pendulous:* hanging downwards. *Erect:* pointing upwards. *Sessile:* stalkless, or apparently so, sitting close upon the body that supports it. The *interspace* always refers to the flowers at half length of the branches at half height of the inflorescence.

Pedicels

The *length* of the pedicels always refers to the ones at half length of the branches at half height of the inflorescence.

Sepals

The *length* of the sepals always refers to the ones at half length of the branches at half height of the inflorescence during anthesis. The terms of describing the *indumentum* of the sepals are the same as for the leaves.

Petals

The *length* of the petals always refers to the ones at half length of the branches at half height of the inflorescence during anthesis. The classification of the *shape* takes into account the state during anthesis and afterwards and is differentiated into four types (Fig. 5):
(a) *Straight during anthesis and afterwards.*
(b) *Recoiled like watchsprings during anthesis and afterwards.*
(c) *Recurved during anthesis, straight afterwards.*
(d) *Recoiled during anthesis, straight afterwards.*

Filaments

The *length* of the filaments always refers to the ones at half length of the branches at half height of the inflorescence during anthesis.

Anthers

The *length* of the anthers always refers to the ones at half length of the branches at half height of the inflorescence during anthesis.

Style

The *length* of the style always refers to the ones at half length of the branches at half height of the inflorescence during anthesis.

Stigmatic complex

The classification of the stigmatic complex follows BROWN & GILMARTIN (1984 b) and SCHILL et al. (1988). Two morphological stigma types are differentiated in *Fosterella*.
Simple-erect: consisting of three free, erect, stylar lobes, each with an introrsely oriented stigmatic line along the distal margin of the lobe (Type I according to BROWN & GILMARTIN 1984 b).
Conduplicate-spiral: consisting of three laminar stylar lobes, each conduplicately folded and with a stigmatic line along the margin of the blade. The three folded lobes are twisted together to yield three spiral stigmatic surfaces, each representing the paired stigmatic margins of a conduplicately folded stylar blade (Type II according to BROWN & GILMARTIN 1984 b).

Capsule

The *form* and *measures* of the capsule always refer to the state immediately before dehiscence.

Seeds

The terms describing the shape of the seeds are used according to STEARN (2004). *Filiform:* slender like a thread. *Clavate:* gradually thickening from a very tapering base.

a   F. spectabilis
    (Peters 06.0045):
    Petals straight
    during anthesis
    and afterwards.

b   F. windischii
    (Krantz 213):
    Petals recoiled like
    watchsprings
    during anthesis
    and afterwards.

c   F. penduliflora
    (Peters 06.0039):
    Petals recurved
    during anthesis,
    straight
    afterwards.

d   F. weberbaueri
    (Krömer 7286):
    Petals recoiled
    during anthesis
    but straight
    afterwards. Photo:
    Timm Stolten.

Fig. 5: Exemplification of petal shape types within the genus *Fosterella*.

Chromosome numbers and genome size determinations

Due to the fact that chromosome counts in bromeliads are notoriously difficult (high numbers, small chromosome sizes, presence of numerous raphid crystals in the cells), my own preliminary attempts to count chromosomes from *Fosterella* root tip preparations gave only ambiguous results.

But then, more or less accidentally, in an attempt to establish molecular microsatellite markers for *Fosterella*, the tested markers provided an interesting result. Thus, all accessions of *F. penduliflora* consistently showed three to five different-sized PCR-amplified bands per locus on a sequencing gel. All other *Fosterella* species included in the test set showed only one or two bands (= microsatellite alleles) at these loci, as would be expected for diploids (BRAUER 2006). This could indicate a tetra- or even hexaploid status of *F. penduliflora*. To further test this expectation, the genome sizes of seven selected *Fosterella* species (Tab. 1) were measured by flow cytometry, using *Pisum sativum* and *Glycine max* as standards. These analyses were kindly performed by Dr. Frank Blattner at the IPK in Gatersleben:

Tab. 1: *Fosterella* specimens subjected to flow cytometry

F. albicans	Rex & Schulte 171002-6
F. gracilis	Rex & Schulte 281002-3
F. micrantha	Welz 3124
F. penduliflora	Ibisch & Vásquez 00.0035, Vásquez 3730, Vásquez 3817, Vásquez 4051, Vásquez 3762, Ibisch 98.0125, Vásquez 3406, Vásquez & Quispe 3407, Ibisch 02.0006, Vásquez 3724 a
F. rusbyi	Rex & Schulte 261002-4
F. weddelliana	Vásquez 3636
F. yuvinkae	Reichle P-SR1, Vásquez 4510

2.3 Evaluation of biogeographical data

For every voucher and living plant, all informations about the collector, collection number and date, geographical coordinates, altitude, locality and habitat, state of bloom and preserving institutions were recorded in a database which was used for the analysis of distribution patterns. Where no geographical coordinates of the collection site were available, they were assigned to the respective specimen by search for the localities on local maps, taking into account roads, rivers, altitude and other relevant features (subsequently added coordinates are put in parentheses). Regarding the distribution of *Fosterella micrantha* in Mexico, some additional collection sites were taken from ESPEJO-SERNA et al. (2004, 2005, 2007). Information describing the habitats, in particular the accompanying species, were taken from notes on herbarium vouchers as indicated. For describing Bolivian ecoregions, the terminology of IBISCH et al. (2004) was used, other ecoregions were named after the NATIONAL GEOGRAPHIC SOCIETY & WWF (2009). Distribution maps were accomplished with the kind assistance of Christoph Nowicki, University of Applied Sciences, Eberswalde.

2.4 Reconstruction of character evolution

In order to reconstruct character evolution within *Fosterella*, the states of ten selected morphological characters were encoded. All character states included in this analysis had been evaluated during the present work on the revision. In cooperation with Dr. Katharina Schulte, Research Institute Senckenberg, Frankfurt/M., the character states were mapped onto the strict consensus tree resulting from parsimony analysis of the combined data matrix of four chloroplast loci, excluding the coded indels (REX et al. 2009) using the software MacClade 4.06 (MADDISON AND MADDISON 2003).

3 Results

3.1 *Fosterella* L.B. Sm.

Fosterella L.B. Sm., Phytologia 7: 171. 1960.
Type species: *Fosterella micrantha* (Lindl.) L.B. Sm.

Etymology
The genus is dedicated to Mulford B. Foster (1888–1978), co-founder and first president of the Bromeliad Society International, who has discovered and collected an extraordinary large number of new species of Bromeliaceae.

Description
Plants terrestrial or saxicolous, perennial, reproducing by suckers, monocarpic, acaulescent or slightly caulescent, 25–200 cm high. **Leaves** 5–30, spirally arranged, rosulate, forming a flat to more or less upright rosette but never a distinct cup. **Sheaths** broadly ovate to triangular, 10–80 mm high, 15–70 mm wide, entire or slightly serrate, green, whitish or reddish, glabrous or abaxially slightly lepidote, somewhat succulent. **Blades** simple, linear, narrowly-triangular or narrowly to broadly (ob-)lanceolate, rarely petiolate, acuminate to acute or cuspidate, 10–100 cm long and 1–8 cm wide, thin to slightly succulent towards the base, entire or serrate towards the base, rarely undulate, adaxially glabrous or sparsely lepidote towards the base, abaxially with more or less dense indumentum of peltate or stellate trichomes. **Inflorescence** terminal, pedunculate, indeterminate, racemose to paniculate with lateral branches up to 3rd order, erect, 5–80 cm long and 5–30 cm wide, axes green or reddish, glabrous, villous or arachnoid, rarely glaucous. **Peduncle** erect, 10–120 cm long, 1–10 mm in diameter, green or reddish, glabrous, villous or arachnoid, rarely glaucous. **Peduncle bracts** more or less appressed, narrowly triangular, acuminate, 1–25 cm long, lower ones often subfoliaceous, much shorter than the internodes, sometimes imbricate, entire or slightly serrate, green, reddish or stramineous, abaxially glabrous, villous, arachnoid or lepidote. **Primary bracts** more or less appressed to the lateral branches, narrowly triangular, acuminate, 5–50 mm long, shorter or longer than the sterile base of the branches, entire or obscurely serrate, green, reddish or stramineous, abaxially glabrous, villous, arachnoid or lepidote. **Branches** up to 30, ascending or inclined, arcuate or straight, 3–30 cm long, bearing 5–80 flowers. **Secondary branches** up to 15, ascending, 2–15 cm long, bearing 4–25 flowers. **Floral bracts** broadly ovate to triangular, acute, 1–12 mm long, shorter or longer than the pedicels, entire, green, reddish or stramineous, abaxially glabrous, villous, arachnoid or lepidote. **Flowers** spreading or secund, erect or pendulous, sessile or pedicelled, bisexual, actinomorphic, 1–15 mm apart. **Pedicels** to 15 mm

long. **Sepals** 3, free, convolute, broadly ovate, obtuse, distinctly shorter than the petals, 1–9 mm long, green, reddish, or stramineous, glabrous, villous, arachnoid, rarely glaucous. **Petals** 3, free, narrowly ovate, obtuse, not appendaged, 3–24 mm long, white, pale greenish or rose, yellow or coral-red, glabrous, straight, recurved during anthesis and straight afterwards or recoiled like watchsprings during anthesis and afterwards. **Stamens** 6, in two whorls, filaments separate from each other, the inner adnate to the base of the petals, exceeding the corolla, 3–15 mm long; anthers basifixed, linear, coiled at anthesis, 1–3 mm long. **Pollen** oblat, monocolpat, 33.6–36.6 µm × 18.1–26.9 µm, brochi ca. 1 µm, muri 1.0–1.3 µm, exine 1.0–1.8 µm, foveolate (EHLER & SCHILL 1973). **Ovary** superior, narrowly ovoid, 4 mm long, trilocular, apocarpous, deeply grooved between the carpels, axile placentae extending to most of the height of the locules. **Style** apical, 1.5–15 mm long, white. **Stigmatic complex** simple-erect or conduplicate-spiral, papillose, margins curled. **Fruit** capsular, narrowly ovoid to globose, 3–8 mm long, 2–4 mm wide, dehiscing from the top: septicidal, slightly loculicidal. **Seeds** brown with white membranous appendages and a narrow dorsal wing connecting them, clavate or filiform to slightly fusiform and bicaudate, 2–4 mm long.

Chromosome number $2n$ = 50, but polyploidy seems to be relatively common.
Flow cytometry yielded the following results: for all *F. penduliflora* samples genome sizes of approximately 1.1 to 1.2 pg were determined, whereas the two investigated *F. yuvinkae* accessions showed values of 0.7 to 0.8 pg. The five remaining species *F. albicans*, *F. gracilis*, *F. micrantha*, *F. rusbyi* and *F. weddelliana* yielded values between 0.35 and 0.4 pg. Interestingly, these values roughly correspond to a ratio of 3:2:1, which would be consistent with a hexaploid status of *F. penduliflora*. Information on individual chromosome numbers of *Fosterella* species from the literature is summarised below (Tab. 2).

Tab. 2: Reported chromosome numbers in *Fosterella* species.

F. albicans	$2n$ = 75, triploid (BLATTNER, pers. comm.).
F. gracilis	$2n$ = 50, diploid (BLATTNER, pers. comm.).
F. micrantha	$2n$ = 50, diploid (BLATTNER, pers. comm.).
F. penduliflora	$2n$ = 50, diploid (BROWN & GILMARTIN 1986), $2n$ = 100, tetraploid (DELAY 1974 a,b), $2n$ = 150, hexaploid (MARCHANT 1967, BROWN & GILMARTIN 1986, BLATTNER, pers. comm.).
F. rusbyi	$2n$ = 50, diploid (BROWN & GILMARTIN 1984 a, 1989 a).
F. villosula	$2n$ = 150, hexaploid (BROWN et al. 1997).
F. weberbaueri	$2n$ = 50, diploid (BROWN & GILMARTIN 1984 a).
F. weddelliana	$2n$ = 50, diploid (BLATTNER, pers. comm.).
F. yuvinkae	$2n$ = 100, tetraploid (BLATTNER, pers. comm.).

Distribution and ecology

The genus is distributed along the eastern slopes of the Andes in central South America from southern Peru in the North (Dpto. Ucayali, approx. 05° 02' S, 75° 30' W) to northern Argentina in the South (Prov. Tucuman, approx. 27° 06' S, 65° 30' W), and western Brazil (Estado Mato Grosso, approx. 15° 30' S, 55°45' W) and northern Paraguay (Dpto. Amambay, approx. 22° 40' S, 56° 05' W) in the East. Apart from this main distribution area with its centre in Bolivia, there are two isolated ranges: *Fosterella batistana* occurs in northern Brazil (Estado Pará, approx. 08° 43' S, 55° 02' W) and *F. micrantha* in Central America from Mexico in the North (Dpto. Oaxaca, approx. 17° 44' N, 96° 19' W) to El Salvador in the South (Dpto. La Libertad, approx. 13° 30' N, 89° 18' W).

Countries:	MEXICO. Estado Guerrero, Veracruz, Oaxaca, Chiapas. GUATEMALA. Dpto. Huehuetenango, San Marcos, Sacatepequez, Retalhuleo, Suchitepequez, Escuintla. EL SALVADOR. Dpto. Chalatenango, La Libertad, Ahuachapan, Cuscatlán, Cabañas. PERÚ. Dpto. Ucayali, Loreto, Huánuco, Pasco, Junín, Cuzco, Ayacucho, Puno. BOLIVIA. Dpto. La Paz, Beni, Cochabamba, Santa Cruz, Chuquisaca, Tarija. ARGENTINA. Prov. Jujuy, Salta, Tucuman. BRAZIL. Estado Mato Grosso, Mato Grosso do Sul, Pará. PARAGUAY. Dpto. Amambay.
Ecoregions:	Tropical to subtropical, evergreen to deciduous forests in Central America: Sinaloan-Dry-Forests, Jalisco-Dry-Forests, Central-American-Pine-Oak-Forests, Southern-Pacific-Dry-Forests, Central-American-Dry-Forests, Veracruz-Moist-Forests, Petén-Veracruz-Moist-Forests, Sierra-Madre-de-Chiapas-Moist-Forests. Dry to humid, evergreen to deciduous lowland forests in central South America: Pre-Andean-Amazon Forests, Sub-Andean-Amazon-Forests, Beni-and-Santa-Cruz-Amazon-Forests, Chiquitano Dry Forests, Cerrado Forests, Ucayali-Moist Forests, Madeira-Tapajós-Moist-Forests, Paraná-Paraíba-Interior-Forests. Dry to humid, evergreen to deciduous montane forests in central South America: Peruvian and Bolivian Yungas, Tucuman-Bolivian-Forests, Inter-Andean-Dry-Forests, Montane Chaco.
Life style & habitats:	Terrestrial or saxicolous on steep slopes, exposed and dry to shady and rather moist, in shrubberies and more or less open forests, secondary habitats along roadsides, riverbanks and small ravines.
Altitude:	100–2750 m.

Taxonomic delimitation and systematic relationships

The genus *Fosterella* L.B. Sm. was established in 1960 by SMITH who split up *Lindmania* Mez into two genera. The group including the generic type, *Lindmania guianensis* (Beer) Mez, was reduced to synonymy of *Cottendorfia* Schult. & Schult.f. and the remainder of *Lindmania* constituted the newly described genus *Fosterella*. Later the genus *Lindmania* was resurrected by SMITH (1986), because he found two relevant indications of generic difference between *Lindmania* and *Cottendorfia*.

In *Fosterella*, with its more or less mesophytic species ranging from Mexico to Argentina and W Brazil, the placentae are axial, extending to most of the height of the locules with numerous ovules. The inner filaments are adnate to the base of the petals, and the anthers basifixed, linear and coiled at anthesis. The leaf blades are rather thin and more or less contracted at the base.

In *Cottendorfia* and *Lindmania*, the placentae are basal and very small with only few ovules. The filaments are all free and the anthers equitant and versatile, stout and straight. The leaf blades are firm and not contracted at the base.

The monotypic genus *Cottendorfia* is restricted to xeric habitats in northeastern Brazil. Its sepals are cochleate with the anterior sepal overlapping the two posterior ones and the flowers are unisexual and dioecious. However, in *Lindmania*, with its species of the Guayana Highland, the sepals are convolute with the left side of each overlapping the right side of the next one and the flowers are bisexual.

According to molecular systematic trees, *Fosterella*, *Cottendorfia* and *Lindmania* are not closely related with each other, but belong to three different subfamilies: *Fosterella* to the Pitcairnioideae *s.str.*, *Cottendorfia* to the Navioideae and *Lindmania* to the Lindmanioideae (GIVNISH et al. 2007, REX et al. 2009).

Besides *Fosterella*, the Pitcairnioideae *s.str.* comprise four other genera, i.e., *Deuterocohnia*, *Dyckia*, *Encholirium* and *Pitcairnia*. These genera differ from *Fosterella* in the following character states:

In *Deuterocohnia* Mez, the petals each bear a basal appendage, whereas in *Fosterella* the petals are not appendaged. The xerophytic species of *Deuterocohnia* are distributed along the southern Andes from Peru to Chile, Argentina and west Brazil.

In *Dyckia* Schult. & Schult.f., the bases of the filaments are either fused to a tube or they are all adnate to the petals, whereas in *Fosterella* only the three inner petals are adnate to the base of the petals. The species of *Dyckia* are distributed in Brazil, Uruguay, Paraguay, Argentina and Bolivia.

The species of *Encholirium* Mart. ex Schult. & Schult.f. have orbicular, reniform or lenticular seeds with a circular to falciform appendage, whereas in *Fosterella* the seeds usually are filiform with bicaudate appendages (exception: *Fosterella nicoliana* has clavate seeds). The species of *Encholirium* are distributed in northeast Brazil.

In *Pitcairnia* L'Hér the seeds are unappendaged at maturity and the corolla usually is zygomorphic, whereas in *Fosterella* the seeds are appendaged and the corolla is actinomorphic. *Pitcairnia* species are distributed from Mexico and the West Indies to Argentina and Brazil.

3.2 Key to the species of the genus *Fosterella* (based on IBISCH et al. 2008)

A	Petals straight or lily-like recurved during anthesis, but becoming straight afterwards; leaf blades entire .... **Subkey I**	23
B	Petals recoiled like watchsprings during anthesis and afterwards; leaf blades entire or serrate .... **Subkey II**	24

Subkey I

1	Flowers bright red, reddish to coral-orange; petals > 20 mm long .... *F. spectabilis*	144
1*	Flowers white, cream or yellow; petals much shorter than 20 mm 2	
2	Flowers and fruits (sub)secund 3	
2*	Flowers and fruits spreading 12	
3	Leaf blades very sparsely lepidote to glabrescent beneath, to 3 cm wide, narrowed towards the base, inflorescence racemose or paniculate, with lateral branches of 1st and rarely 2nd order; axes slender, green, sparsely arachnoid; flowers pendulous; floral bracts sparsely arachnoid .... *F. batistana*	39
3*	Leaf blades distinctly covered by trichomes beneath, 1–8 cm wide, rarely narrowed towards the base, but not petiolate 4	
4	Inflorescence glabrous 5	
4*	Inflorescence covered by trichomes throughout, often also the flowers 9	
5	Leaf blades lepidote beneath, bearing peltate trichomes 6	
5*	Leaf blades tomentose beneath, bearing stellate trichomes 7	
6	Leaf blades densely appressed lepidote beneath by peltate trichomes with fimbriate margin, to 20 cm long and 2 cm wide; primary bracts longer than the sterile base of the branches; pedicels to 2 mm long; sepals 1.5 mm long; petals 4–5 mm long .... *F. elviragrossiae*	61
6*	Leaf blades scattered lepidote beneath by peltate trichomes with dentate margin, to 40 cm long and 8 cm wide, sometimes reddish; primary bracts shorter than the sterile base of the branches; pedicels 2–6 mm long; sepals 2–3 mm long; petals 7–10 mm long .... *F. penduliflora*	105
7	Leaf blades sparsely tomentose beneath; petals yellow; inflorescence axes glaucous, branches curved ascending .... *F. gracilis*	70
7*	Leaf blades densely tomentose beneath; petals white 8	
8	Leaf blades to 50 cm long; sepals 2–3 mm long; petals 7 mm long ..... *F. hatschbachii*	78
8*	Leaf blades 20–35 cm long; sepals 3–4 mm long; petals 7–9 mm long ...... *F. yuvinkae*	172

9	Inflorescence axes arachnoid; fruits globose; seeds clavate; leaf blades scattered lepidote beneath, glabrescent; floral bracts 1–2 mm long; sepals 2–3 mm long; petals 5–6 mm long .... *F. nicoliana*	97
9*	Inflorescence axes villous; fruits ovoid; seeds filiform **10**	
10	Leaf blades light green beneath; peduncle, inflorescence axes and floral bracts sparsely villous; sepals glabrous .... *F. micrantha*	90
10*	Leaf blades green or reddish beneath; peduncle and inflorescence axes densely villous; floral bracts and sepals villous **11**	
11	Floral bracts 3–5 mm long; sepals 2–3 mm long .... *F. christophii*	52
11*	Floral bracts 7 mm long; sepals 4–5 mm long .... *F. villosula*	152
12	Flowers suberect; floral bracts whitish; sepals white, petals recoiled at anthesis, but becoming straight afterwards .... *F. weberbaueri*	157
12*	Flowers erect; floral bracts green; sepals green **13**	
13	Sepals 2–3 mm long; peduncle, inflorescence axes and sepals glabrous; petals recurved, 5–7 mm long; leaf blades scattered lepidote to glabrescent beneath ...... *F. chaparensis*	48
13*	Sepals 6–8 mm long; peduncle, inflorescence axes and sepals arachnoid; petals straight, forming a tube, 8–12 mm long; leaf blades densely appressed lepidote beneath .... *F. floridensis*	65

Subkey II

1	Flowers erect or spreading **2**	
1*	Flowers pendulous and secund **7**	
2	Leaves few and small, broadly lanceolate, 10–40 cm long, 10–40 mm wide, scattered lepidote to glabrescent beneath, entire, strongly nerved; inflorescence racemose or with only very few, short primary branches, axes sparsely arachnoid; peduncle bracts small and remote .... *F. aletroides*	35
2*	Leaves long and linear or narrowly (ob)lanceolate, 30–100 cm long, with a dense layer of trichomes beneath **3**	
3	Plant acaulescent **4**	
3*	Plant caulescent **6**	
4	Peduncle and inflorescence axes slightly arachnoid, glabrescent; leaf blades to 2 cm wide; densely appressed lepidote beneath, entire (rarely some minute spines at the base); floral bracts 1 mm long; sepals 1.5 mm long; petals 4 mm long .. *F. robertreadii*	123
4*	Peduncle and inflorescence axes densely arachnoid **5**	

5	Leaf blades 2–3 cm wide, covered by a thick layer of interwoven trichomes beneath, serrate toward the base; inflorescence rather dense; floral bracts 4–5 mm long; sepals 3–4 mm long; petals 6–8 mm long .... *F. albicans*	27
5*	Leaf blades to 1–1.5 cm wide, densely appressed lepidote beneath, entire; inflorescence rather lax; floral bracts 3–4 mm long; sepals 2–3 mm long; petals 5 mm long .... *F. pearcei*	101
6	Peduncle and inflorescence axes densely arachnoid; flowers subsessile; floral bracts 5–7 mm long, sparsely arachnoid; sepals 5 mm long, sparsely arachnoid; petals 7–8 mm long, pale greenish; leaf blades 30–60 cm long, 2–3 cm wide, serrate towards base, densely appressed lepidote beneath .... *F. caulescens*	43
6*	Peduncle and inflorescence axes slightly arachnoid; flowers sessile; floral bracts 2 mm long, sparsely villous; sepals 3 mm long, sparsely villous to glabrescent; petals 5 mm long; leaf blades 20–30 cm long, to 1.7 cm wide, serrate towards base, densely appressed lepidote beneath .... *F. rexiae*	119
7	Leaf blades serrate 8	
7*	Leaf blades entire or with some inconspicuous spines at the base 12	
8	Leaf blades to 1.2 cm wide, densely appressed lepidote beneath, linear, to 60 cm long, strongly involute, serrate towards base; peduncle glabrous and with long, foliaceous bracts; inflorescence axes glabrous; flowers secund; floral bracts 1 mm long; pedicels to 5 mm long; sepals 1.5 mm long; petals 3 mm long .... *F. graminea*	74
8*	Leaf blades 1.5–6 cm wide, covered by a thick layer of interwoven trichomes beneath 9	
9	Plant acaulescent 10	
9*	Plant caulescent 11	
10	Leaf blades to 6 cm wide, distinctly serrate at base, undulate; plant up to 200 cm high; inflorescence branches arcuate ascending; petals frequently greenish or rose *F. rusbyi*	132
10*	Leave blades 1.5–3.5 cm wide, obscurely serrate; plant up to 50 cm high; inflorescence branches erect spreading; petals white .... *F. vasquezii*	148
11	Leaf blades to 2 cm wide; peduncle bracts slightly serrate; floral bracts 4–7 mm long; sepals 5 mm long; petals 8 mm long ... . *F. cotacajensis*	56
11*	Leaf blades 2–3.5 mm wide; peduncle bracts distinctly serrate; floral bracts 2–3 mm long; sepals 2–3 mm long; petals 4–5 mm long .... *F. weddelliana*	162
12	Leaf blades scattered lepidote to glabrescent beneath; blades narrowly lanceolate, 35–40 cm long and 2–3 cm wide, narrowed toward the base; peduncle and inflorescence axes glabrous .... *F. schidosperma*	139
12*	Leaf blades covered by a dense layer of trichomes beneath 13	

13	Plant caulescent, up to 60 cm high; leaf blades densely appressed lepidote beneath; dimorphic foliation: lower leaves surrounding the stem, narrowly triangular, gradually merging into the petiolate rosette leaves; peduncle frequently reddish, glabrous; peduncle bracts appressed lepidote .... ***F. heterophylla***	82
13*	Plant acaulescent 14	
14	Leaf blades not or only slightly narrowed towards the base; pedicels to 5 mm long; floral bracts much shorter than pedicels; leaf blades covered by a thick layer of interwoven trichomes beneath; peduncle bracts 4–7 cm long; petals 5–6 mm long .... ***F. rojasii***	128
14*	Leaf blades narrowed towards the base, more or less petiolate 15	
15	Leef blades petiolate, to 3.5 cm wide, entire, densely appressed lepidote beneath; primary bracts 20–30 mm long; peduncle bracts 5–8 cm long, appressed lepidote, glabrescent; sepals 1–2 mm long; petals 3–5 mm long .... ***F. petiolata***	114
15*	Leaf blades narrowed towards the base, almost petiolate, to 3 cm wide, entire or with some inconspicuous spines at the base; primary bracts to 10 mm long 16	
16	Leaf blades to 3 cm wide, densely appressed lepidote beneath; peduncle bracts appressed lepidote, equalling the internodes; inflorescence with more than 15 branches, branches up to 5 cm long .... ***F. kroemeri***	86
16*	Leaf blades 1.5–2.5 cm wide, covered by a thick layer of interwoven trichomes beneath; peduncle bracts interwoven lepidote, always longer than the internodes; inflorescence with less than 15 branches, branches longer than 5 cm ..... ***F. windischii***	168

3.3 Taxonomic treatment

Fosterella albicans (Griseb.) L.B. Sm., Phytologia 7: 171. 1960. Fig. 6–10.

Basionym:	*Cottendorfia albicans* Griseb., Abh. Königl. Ges. Wiss. Göttingen, Math.-Phys. Kl. 24: 330. 1879. TYPE: Argentina. Prov. Salta: Dpto. Orán, valley of Río Seco below San Andres, 17–24 Sept. 1873, *P.G. Lorentz & G. Hieronymus 502* [Lectotype: GOET!; Isotypes: B!, CORD (phot.!), NY!].
≡	*Lindmania albicans* (Griseb.) Mez in C.DC., Monogr. Phan. 9: 537. 1896.
=	*Fosterella fuentesii* Ibisch, R. Vásquez & E. Gross, Selbyana 23 (2): 207. 2002. TYPE: Bolivia. Dpto. Santa Cruz: Prov. Florida, Municipio de Pampagrande, Manzanillares, east of Valle Hermoso, 17° 53' 59.1" S, 64° 10' 13.6" W, 1820 m, 10 Aug. 2001, *A. Fuentes, A. Araúz & I. Rivera 3176* [Holotype: LPB!; Isotype: FR!, SEL!, USZ!].

Comment on type

The type locality of *Cottendorfia albicans* Griseb. is situated at approx. 23° 05' W, 64° 37' S. The specimen *P.G. Lorentz & G. Hieronymus 502* was designated as lectotype in PETERS et al. (2008 b), for the following reasons: GRISEBACH (1879) did not choose a holotype, when he described *Cottendorfia albicans*, but just gave a short information about the origin of the described specimen and stated explicitly that it was collected by Lorentz and Hieronymus. BAKER (1889) cited two specimens, *Lorentz & Hieronymus 288* and *502*, which obviously were available to Grisebach. He listed both as types, which was adopted in following publications dealing with this species (MEZ 1896, SMITH 1934 b). Despite of the different numbering, the two collections were most likely made at the same date and locality: In GOET, three specimens collected by *Lorentz & Hieronymus* are preserved: two specimens numbered as *288*, both consisting of a simple rosette only, and a third, no. *502*, containing two inflorescences, which – according to the length of scapes – fit to the others very well. All three sheets bear labels from the same hand, indicating the same origin and date of collection. SMITH & DOWNS (1974) designated *Lorentz & Hieronymus 502* (GOET) as holotype, obviously because it contains the generative parts of the plant, and several original pencil drawings of floral elements. Nevertheless, according to the ICBN, Art. 9.9 (NEILL et al. 2006), this specimen should be considered as lectotype. Since the lectotype is fragmentary because of the missing leaves, the other specimens, *Lorentz & Hieronymus 288* [B! (st.), BA!, CORD (phot.!) (st.),GOET! (st.), NY! (st.), US! (st.)], some of them containing both basal leaves and inflorescence, represent most beneficial complement.

Etymology

The epithet refers to the white colour of the petals.

Description

Plant subcaulescent, up to 100 (180) cm high. **Leaves** many, 20–30, forming an open, more or less upright rosette. **Sheaths** 50–70 mm wide, entire, whitish, glabrous. **Blades** narrowly oblanceolate, acuminate, narrowed towards the base, 40–100 cm long and 2–3 (4) cm wide, succulent and serrate towards the base, adaxially sparsely lepidote, abaxially covered by a thick layer of interwoven, peltate trichomes with fimbriate margin. **Peduncle** 50–120 cm long, 5–8 mm in diameter, green, densely arachnoid. **Peduncle bracts** 4–12 cm long, much longer than the internodes, entire, abaxially covered by a layer of interwoven trichomes. **Inflorescence** paniculate with branches up to 2nd order, 20–50 cm long and 8–10 cm wide, axes green, arachnoid. **Primary bracts** 20–30 mm long, longer than the sterile base of the branches, entire, abaxially arachnoid. **Branches** 8–20, inclined, arcuate, 3–18 cm long, bearing up to 50 flowers. **Secondary branches** 2–5 cm long, bearing up to 10 flowers. **Floral bracts** 4–5 mm long, longer than the pedicels, entire, abaxially arachnoid. **Flowers** spreading, suberect, subsessile, 1–5 mm apart. **Pedicels** to 1.5 mm long. **Sepals** 3–4 mm long, green, sparsely arachnoid, glabrescent. **Petals** 6–8 mm long, white, recoiled like watchsprings during anthesis and afterwards. **Filaments** 3 mm long. **Anthers** 1.5 mm long. **Style** 1.5 mm long. **Stigmatic complex** simple-erect. **Capsule** ovoid, 5–6 mm long, 3 mm wide. **Seeds** filiform, bicaudate, 3–4 mm long.

Specimens seen

BOLIVIA: **Dpto. La Paz:** Prov. Larecaja: Valley of Consata, 15° 24' 36" S, 68° 32' 24" W, 1800 m, 8 Aug. 2002, *Nowicki & Deichmann 2359* [FR!]; ibid.: Valley of Consata, 15° 24' 36" S, 68° 32' 24" W, 1800 m, 8 Aug. 2002, *Nowicki & Deichmann 2360* [FR!]. Prov. Yungas: 23 km from Coroico to Unduavi, behind Yolosa, 16° 14' 46" S, 67° 47' 08" W, 1975 m, 21 Oct. 2006, *Peters 06.0100* [LPB!]; ibid.: 23 km from Coroico to Unduavi, behind Yolosa, 16° 14' 46" S, 67° 47' 08" W, 1975 m, 21 Oct. 2006, *Peters 06.0101* [SEL!]; ibid.: 23 km from Coroico to Unduavi, behind Yolosa, 16° 14' 46" S, 67° 47' 08" W, 1975 m, 21 Oct. 2006, *Peters 06.0102* [FR!; BGHD!]; ibid.: Río Coroico below Coroico, (16° 04' 48" S, 67° 42' 00" W), 1550 m, 15 July 1989, *Kessler & Kelschebach 146* [GOET!, LPB!]. Prov. Sud Yungas: 84 km from Coroico to Chulumani, 2650 m, 28 Nov. 1995, *Gouda 95-91 b* [U!]; ibid.: Road from Coroico to Chulumani, next to Puente. Villa, (16° 24' S, 67° 30' W), 2450 m, 29 Nov. 1995, *Gouda 95-95 c* [FR!, U!; BGB!]; ibid.: 39 km from Plazuela to La Paz, 16° 38' S, 67° 34' W, 2150 m, 29 Sept. 1995, *Kessler 5664*, [LPB! (st.)]. **Dpto. Cochabamba:** Prov. Chapare: Cordillera Mosetenses, 16° 14' S, 66° 25' W; 1550 m, 2003, *Kessler 13251* [FR!, GOET!; BGGÖ!]; ibid.: Pucu Mayo, 9 km below Corani Pampa, (17° 12' S, 65° 42' W), 1750 m, 23 June 1996, *Wood & Ritter 11236* [K!]; ibid.: 90 km from Cochabamba to Villa Tunari, Locotal, between Corani and Naranjitos, 17° 11' 24" S, 65° 48' 24" W, 1800 m, 22 Oct. 2006, *Peters 06.0103* [LPB!]; ibid.: 90 km from Cochabamba to Villa Tunari, between Corani and Naranjitos, 17° 11' 24" S, 65° 48' 24" W, 1800 m, 22 Oct. 2006, *Peters 06.0104* [SEL!; BGHD!]; ibid.: Road from Villa Tunari to Cochabamba, 5 km before Río Vinto, 17° 11' S, 65° 49' W, 1800 m, 17 Oct. 2002, *Vásquez 4626 b* [FAN!]; ibid.: Road from Villa Tunari to Cochabamba, 5 km before Río Vinto, 17° 11' 40" S, 65° 48' 40" W, 1760 m, 17 Oct. 2002, *Rex & Schulte 171002-6* [FR!, LPB! (st.), SEL! (st.); BGHD!]; ibid.: Between Cochabamba and Villa Tunari; next to Sacaba, 17° 42' S, 65° 48' W, 1750 m, 5 Feb. 1999, *Vásquez 3155* [SEL!]; ibid.: Between Cochabamba and Villa Tunari, (17° 10' 12" S, 65° 48' 00" W), 1750 m, 2. Sept. 1999, *Vásquez 3317* [FR!, GOET!, LPB!]; ibid.: Next to Sacaba, (18° 04' S, 66° 23' W), 2000 m, 27 Aug. 1921, *Steinbach 5699* [LIL!]. Prov. Carrasco: 143 km on the old road from Cochabamba to Villa Tunari, 17° 07' S, 65° 34' W, 1300 m, 27 Aug. 1996, *Kessler 7832*, [SEL! (st.)]. **Dpto. Santa Cruz:** Prov. Florida: El Sillar, 1350 m, 5 Nov. 2000, *Vásquez 3796 a* [USZ!; FAN!]; ibid.: Municipio de Pampagrande, Manzanillares, east of Valle Hermoso, 17° 53' 59" S, 64° 10' 13" W, 1820 m, 10 Aug. 2001, *Fuentes 3176* [FR!, LPB!, SEL!]; ibid.: Municipio de Pampagrande, Sierra Racete, Manzanillares, east of

Valle Hermoso, 18° 08' 55" S, 63° 55' 48" W, 1530 m, 1 Oct. 2006, *Peters 06.0002* [SEL!; BGHD!]; ibid.: Municipio de Pampagrande, Sierra Racete, Manzanillares, east of Valle Hermoso, 18° 08' 55" S, 63° 55' 48" W, 1530 m, 1 Oct. 2006, *Peters 06.0003* [LPB!; BGHD!]; ibid.: Municipio de Pampagrande, Sierra Racete, Manzanillares, east of Valle Hermoso, 18° 08' 55" S, 63° 55' 48" W, 1530 m, 1 Oct. 2006, *Peters 06.0004* [FR!]; ibid.: Municipio de Pampagrande, Sierra Racete, Manzanillares, east of Valle Hermoso, 18° 08' 55" S, 63° 55' 48" W, 1530 m, 1 Oct. 2006, *Peters 06.0005* [BGHD!]; ibid.: Municipio de Pampagrande, Sierra Racete, Manzanillares, east of Valle Hermoso, 18° 08' 55" S, 63° 55' 48" W, 1530 m, 1 Oct. 2006, *Peters 06.0006* [BGHD!]; ibid.: 40 km S of Samaipata, 18° 25' S, 63° 50' W, 1700 m, 28 Dec. 1998, *Ibisch 98.0204* [FR!, LPB!]; Between Samaipata and Santa Cruz, 1200 m, *Fuchs & Naranja s.n.* [SEL!; MSBG!]. Prov. Vallegrande: 12 km from Loma Larga towards Masicurí, 18° 47' S, 63° 57' W, 1250 m, 21 May 1996, *Kessler 5973* [LPB! (st.), SEL! (st.)]; ibid.: 4 km from Loma Larga to Masicurí, 18° 47' S, 63° 53' W, 1850 m, 4 June 1996, *Kessler 6303* [LPB!, SEL!]; ibid.: Road from Valle Grande to Masicuri, next to Loma Larga, 18° 46' 22" S, 63° 53' 23" W, 1875 m, 2 Oct. 2006, *Peters 06.0010* [LPB!; BGHD!]; ibid.: Road from Valle Grande to Masicuri, next to Loma Larga, 18° 46' 22" S, 63° 53' 23" W, 1875 m, 2 Oct. 2006, *Peters 06.0011* [SEL!]; ibid.: Road from Valle Grande to Masicuri, next to Loma Larga, 18° 46' 22" S, 63° 53' 23" W, 1875 m, 2 Oct. 2006, *Peters 06.0012* [FR!]; ibid.: Road from Valle Grande to Masicuri, next to Loma Larga, 18° 46' 22" S, 63° 53' 23" W, 1875 m, 2 Oct. 2006, *Peters 06.0013* [FR!]; ibid.: Road from Valle Grande to Masicuri, next to Loma Larga, 18° 46' 22" S, 63° 53' 23" W, 1875 m, 2 Oct. 2006, *Peters 06.0014* [SEL!; BGHD!]; ibid.: Road from Valle Grande to Masicuri, next to Loma Larga, 18° 46' 56" S, 63° 52' 44" W, 1710 m, 2 Oct. 2006, *Peters 06.0015* [LPB!]. **Dpto. Chuquisaca:** Prov. Jaime Mendoza: Comunidad Corey, 19° 20' S, 64° 04' W, 1500 m, 27 Aug. 1999, *Gonzales & Huaylla 2991* [LPB!]; ibid.: 68 km from Monteagudo to Padilla, 19° 34' S, 64° 09' W, 1400 m, 2 July 1995, *Kessler 4986* [SEL!]; ibid.: 96 km from Monteagudo to Padilla, 19° 32' S, 64° 10' W, 1500 m, 2 July 1995, *Kessler 5033*, [SEL! (fr.)]; ibid.: 62 km from Padilla to Monteagudo, (19° 43' 00" S, 64° 01' 48" W), 1300 m, 1 Oct. 1982, *Beck & Libermann 9354* [LPB!, SEL!]; ibid.: Next to Llantoj and next to Río Jantoj, 19° 19' 00" S, 64° 04' 08" W, 1125 m, 10 Oct. 2004, *Gutiérrez 887* [HSB!]. Prov. Hernando Siles: Road from Monteagudo to Padilla, next to Monteagudo, 19° 47' 13", 64° 02' 23" W, 1216 m, 3 Oct. 2006, *Peters 06.0031* [SEL!]; ibid.: Road from Monteagudo to Padilla, next to Monteagudo, 19° 47' 13" S, 64° 02' 23" W, 1216 m, 3 Oct. 2006, *Peters 06.0032* [LPB!]; ibid.: Road from Monteagudo to Padilla, next to Monteagudo, 19° 47' 13" S, 64° 02' 23" W, 1216 m, 3 Oct. 2006, *Peters 06.0033* [FR!]; ibid.: Road from Monteagudo to Padilla, next to Monteagudo, 19° 47' 13" S, 64° 02' 23" W, 1216 m, 3 Oct. 2006, *Peters 06.0037* [LPB!]; ibid.: 61 km from Monteagudo to Padilla, 19° 33' 28" S, 64° 07' 10" W, 1198 m, 4 Oct. 2006, *Peters 06.0044* [LPB!]; ibid.: 64 km from Monteagudo to Padilla, 19° 32' 19" S, 64° 07' 28" W, 1269 m, 4 Oct. 2006, *Peters 06.0049* [SEL!; BGHD!]. Prov. Luis Calvo: Between Incahuasi and Muyupampa, (19° 51' S, 63° 43' W), 1430 m, 22 Apr. 2000, *Vásquez & Rivero 3729 b* [LPB!]. Prov. Sud Cinti: North of Río Santa Martha, 20° 40' 55" S, 64° 19' 29" W, 1400 m, 30 May 1995, *Holst 4837* [HSB!, SEL!]; ibid.: Puca Pampa, Río San Cristobal Lequina, towards Río Alborniyoj, 20° 45' 36" S, 64° 30' 47" W, 1650 m, 9 Oct. 2004, *Lliully 44* [HSB!]; ibid.: 5 km NW from Las Abras, Cañon Pirua, 21° 05' 34" S, 64° 16' 22" W, 1408 m, 12 Oct. 2005, *Lozano & Cardos 1442* [HSB!]; ibid.: Valley of Río Limonal, N of the hill Bufete, 20° 49' 41" S, 64° 22' 37" W, 1820 m 18 May 1995, *Serrano 1348*, [SEL! (st.)]; ibid.: Valley of Río Santa Martha, 20° 42' 41" S, 64° 18' 19" W, 950 m, 25 May 1995, *Holst 1409* [USZ!]. **Dpto. Tarija:** Prov. Arce: Between Emborozu and la Mamora, 6 Oct. 1973, *Legname & Cuezzo 9608 C* [LIL!]; ibid.: Sidras, 22° 12' S, 64° 35' W, 1000 m, 15 May 2001, *Vásquez 4023* [FR!, SEL!]; ibid.: 118 km from Tarija to Bermejo, valley of Río Bermejo, between La Merced and Vaden, 22° 14' 50" S, 64° 35' 24" W, 953 m, 11 Oct. 2006, *Peters 06.0055* [LPB!; BGHD!]; ibid.: 118 km from Tarija to Bermejo, valley of Río Bermejo, between La Merced and Vaden, 22° 14' 50" S, 64° 35' 24" W, 953 m, 11 Oct. 2006, *Peters 06.0056* [SEL!; BGHD!, BGKS!]; ibid.: 118 km from Tarija to Bermejo, valley of Río Bermejo, between La Merced and Vaden, 22° 14' 50" S, 64° 35' 24" W, 953 m, 11 Oct. 2006, *Peters 06.0057* [FR!; BGHD!, BGKS!]. ARGENTINA: **Prov. Jujuy:** Dpto. Ledesma: Parque Nacional Calilegua, 15–20 km NW of Calilegua, 4 km W of Tome de Agua, next to the border of the PN, 23° 39' 26" S 64°46'41" W, 850 m, 8 July 1999, *Tolaba 1832* [MCNS!]; ibid.: Parque Nacional Calilegua, 15–20 km NW of Calilegua, 4 km W of Tome de Agua, next to the border of the PN, 23° 39' 26" S, 64° 46' 41" W, 850 m, 9 July 1999, *Tolaba 1879* [MCNS!]. **Prov. Salta:** Dpto. Santa Victoria: Parque Nacional Baritu, S of Río Prongal towards the founts of La Canaleta and Río Pes, (22° 15' S, 64° 58' W), 25 Sept. 1990, *Novara 10024* [MCNS!]; ibid.: Parque Nacional Baritu, next to Río Lipeo, between Lipeo and the union with Río Sivingal, 1300 m, 15 Sept. 1990, *Novara 9933* [MCNS!]; ibid.: Río Lipeo, 1100 m, 28 Nov. 1971, *Marmol 8755 C* [LIL!]; ibid.: Río

Poróngal, 800 m, 6 Nov. 1978, *Halloy A 460* [LIL!]. <u>Dpto. Orán:</u> Valley of Río Blanco near Río Seco, (22° 48' S, 64° 06' W), Sept. 1873, *Lorentz & Hieronymus 288* [B! (st.), BA!, CORD (phot.!) (st.),GOET! (st.), NY! (st.), US! (st.)]; ibid.: Río Cañas, (22° 48' S, 64° 06' W), 25 Aug. 1944, *Willink s.n.* [LIL!]; ibid.: Arasayal, 2 km behind Finca Yakulica, 690 m, 26 Oct. 1970, *Vervoorst & Cuezzo 7814 C* [LIL!, NY!]; ibid.: Arasayal, between Finca Yakulica and El Angosto del Pescado, 560 m, 23 Sept. 1974, *Meneses & Vervoorst 36* [LIL!]; ibid.: Finca Yakulica, 21 km from Aguas Blancas, (22° 42' S, 64° 24' W), 1050 m, 15 Sept. 1993, *Wasshausen 1971* [K!]; ibid.: Aguas Blancas, Finca El Arasayal, Valley of Nogalar, 800 m, 24 July 1985, *Palací 118* [MCNS!]; Aguas Blancas, Finca El Arasayal, Valley of Nogalar, 600 m, 26 July 1986, *Palací 753* [MCNS!]; ibid.: Finca El Arasayal, Valley of Nogalar, 800 m, 10 Nov. 1979, *Legname 7007* [LIL!]. **Prov. Tucuman:** <u>Dpto. Monteros:</u> Valley of Acheral, (27° 06' S, 65° 30' W), 27 July 1944, *Castellanos s.n.* [LIL!]. WITHOUT LOCALITY: *BIC 234* [SEL!].

Distribution and ecology

Range size: Relatively large, more or less continuous.

Countries: BOLIVIA. Dpto. La Paz, Cochabamba, Santa Cruz, Chuquisaca, Tarija.

ARGENTINA. Dpto. Jujuy, Salta, Tucuman.

Ecoregions: Tucuman-Bolivian-Forests, Yungas.

Life style & habitats: Clearings in semihumid to humid, deciduous to evergreen montane forests. Extensive scrub, rough grassland and near settlements. Terrestrial or saxicolous, on sandy ground, rocks, steep slopes, along roadsides or riverbanks. Scattered to locally common, abundant and forming colonies.

Tucuman-Bolivian-Forest with *Myrcianthes pseudo-mato* (Myrtaceae), *Cedrela lilloi* (Meliaceae) and various pteridophytes, frequent cattle-intervention (*Lliully 44* [HSB]). Evergreen forest with *Capparis* sp. (Capparaceae) and *Chrysophyllum gonocarpum* (Sapotaceae) (*Gutiérrez 887* [HSB]). Secondary, evergreen forest with *Ceiba pentandra* (Bombaceae) (*Kessler 5973* [LPB]). Rocky slope with *Crassula* sp. (Crassulaceae), *Encyclia* sp. (Orchidaceae) and *Cleistocactus* sp. (Cactaceae) (*Nowicki & Deichmann 2359* [FR]). Semideciduous forest with species of Anacardiaceae, Rubiaceae and Maranthaceae, sporadic cattle-intervention (*Holst 4837* [SEL]).

Altitude: 550–2650 m.

Taxonomic delimitation and systematic relationships

According to molecular data (REX et al. 2009), *Fosterella albicans* is very close to *F. caulescens*, *F. kroemeri* and *F. rexiae*. The genetic heterogeneity of the widely distributed *F. albicans* supports the idea of *F. albicans* being a morphologically and ecologically variable species.

Morphologically, *Fosterella albicans* is quite similar to *F. caulescens* – apart from the thicker and denser indumentum of the abaxial leaf surface, the more compact inflorescence, and the less pronounced caulescence in *F. albicans*. Both species have in common a more or less upright rosette with serrate leaf blades, a densely arachnoid inflorescence and subsessile flowers with petals recoiled like watchsprings.

Fosterella fuentesii Ibisch, R. Vásquez & E. Gross has been synonymised under *Fosterella albicans* (Griseb.) L.B. Sm. in PETERS et al. (2008 b), for the following reasons. When *F. fuentesii* was described (IBISCH et al. 2002), it was assumed to be close to *F. petiolata* and was known from the type locality and dried material only. In the meantime, more specimens have been collected and documented in the field during an expedition to the type locality: BOLIVIA. Dpto. Santa Cruz: Prov. Florida, north of Pampa Grande, Sierra Racete, Manzanillares, east of Valle Hermoso, 18° 08' 55" S, 63° 55' 48" W, 1530 m, 1 Oct. 2006, *Peters 06.0002* [SEL!; BGHD!], *Peters 06.0003* [LPB!, BGHD!], *Peters 06.0004* [FR!]. The examination of plenty of material revealed that these specimens have to be named *F. albicans*, which appears to be a more variable taxon than thought before. The subsequent critical recheck of the type material of *F. fuentesii* led to the conclusion that it belongs to *F. albicans* as well. The similarity to *F. petiolata* was striking at first glance because of the distinct petioles, the conspicuous primary bracts, and the rather lax inflorescence. But with the serrate petioles, the thickly lepidote abaxial leaf surface, the villous-arachnoid scape and inflorescence-branches, the size of the floral bracts, sepals and petals, and the habitat, the specimen resembles much more to *F. albicans*.

The specimen *Peters 06.0037* [LPB!] from the Tucuman-Bolivian-Forests in the Dpto. Chuquisaca, Prov. Hernando Siles shows two abnormal characteristics: the leaf blades are not serrate and the inflorescence is only very sparsely arachnoid. Interestingly it was found among a population of plants which definitely have to be identified as *F. penduliflora*. But since all other morphological characters coincide, it is classified under *F. albicans* here.

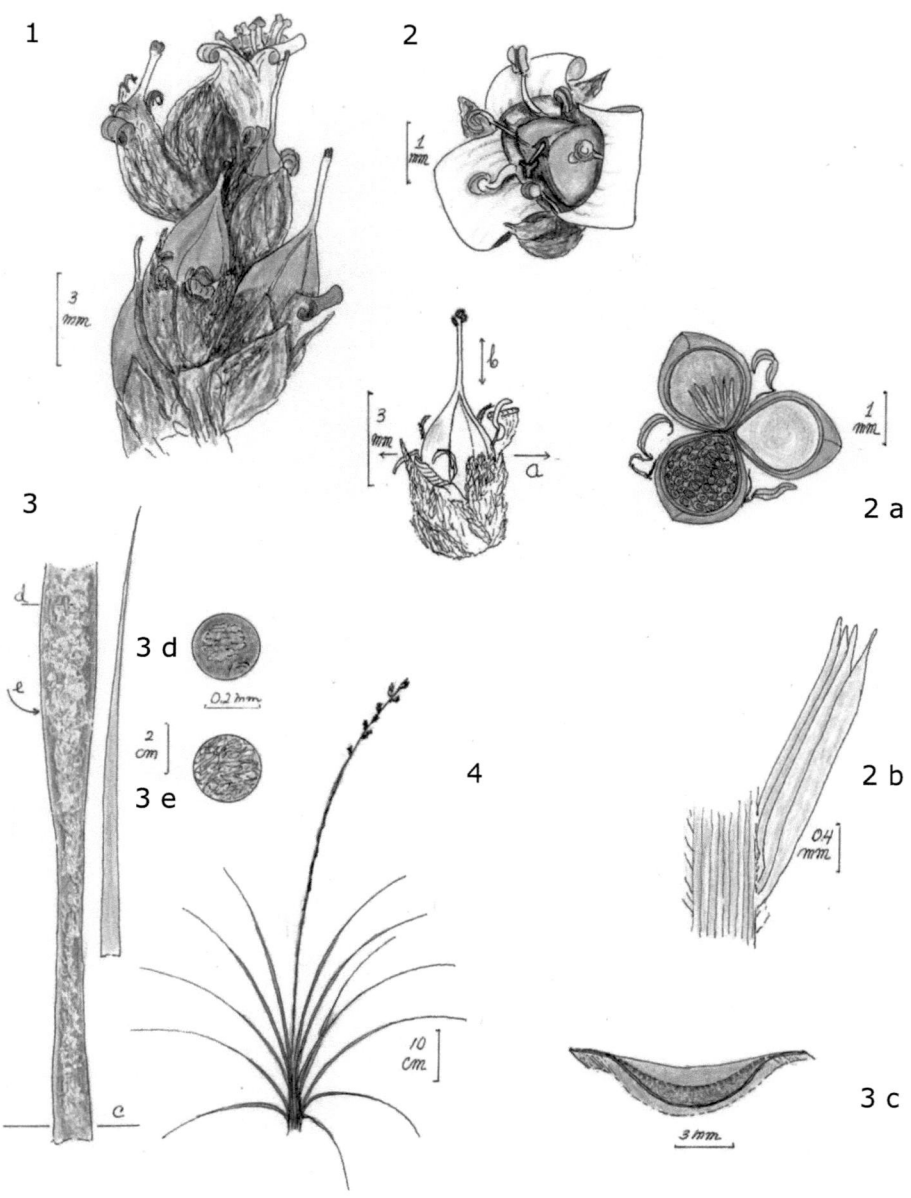

Fig. 6: *Fosterella albicans*. 1 Inflorescence. 2 Flower 2 a Cross section. 2 b Longitudinal section. 3 Leaf. 3 c Cross section at the base of the leaf. 3 d Adaxial leaf surface. 3 e Abaxial leaf surface. 4 Habit. Drawing of *Gouda 95-91 b* by Dr. C.D. Laros. © Botanical Gardens Utrecht.

Fig. 7: Lectotype of *Fosterella albicans* (Griseb.) L.B. Sm. [GOET].

Fig. 8: *F. albicans (Peters 06.0055)* in the natural habitat.

Fig. 9: Flowers of *F. albicans (Peters 06.0006)*.

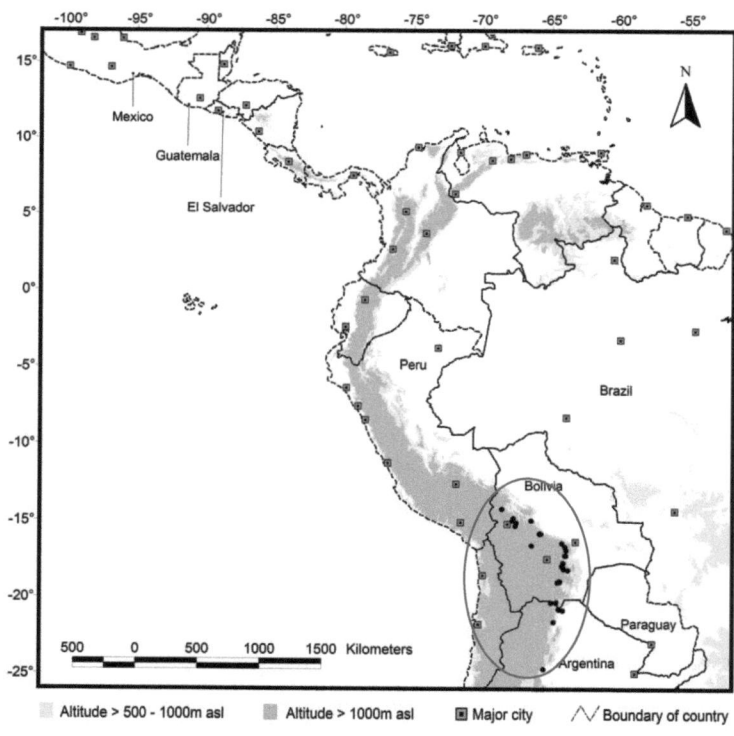

Fig. 10: Distribution of *Fosterella albicans*.

Fosterella aletroides (L.B. Sm.) L.B. Sm., Phytologia 7: 171. 1960. Fig. 11–13.

Basionym: *Lindmania aletroides* L.B. Sm., Contr. U.S. Natl. Herb. 29: 530. 1954. TYPE: Perú. Dpto. Cuzco: Prov. Paucartambo, Sapansachayoc, border of woods, 900 m, 24–30 July 1948, *C. Vargas 7358* [Holotype: US (phot.!)].

Comment on type

According to PERCY NÚÑEZ et al. (2001), Sapansachayoc is a locality in the "Reserva de Biosfera del Manu", situated at an altitude of 1250 m.

Etymology

The epithet means that the species resembles *Aletris* L., Melanthiaceae.

Description

Plant acaulescent, up to 35 cm high. **Leaves** few, 5–8, forming an open, more or less upright rosette. **Sheaths** 15–20 mm wide, entire, whitish, glabrous. **Blades** oblanceolate, more or less petiolate, acuminate, 10–40 cm long and 1–4 cm wide, thin, entire, adaxially glabrous, abaxially scattered lepidote by peltate trichomes with dentate margin, glabrescent. **Peduncle** very slender, 1–2 mm in diameter, green, sparsely arachnoid. **Peduncle bracts** 1–1.5 cm long, equalling the internodes, entire, abaxially scattered lepidote, glabrescent. **Inflorescence** racemose or compound racemose, 5–20 cm long and to 7 cm wide, axes green, sparsely arachnoid. **Primary bracts** 5–10 mm long, longer than the sterile base of the branches, entire, abaxially sparsely stellate lepidote, glabrescent. **Branches** up to 12, ascending, straight, to 6 cm long, bearing up to 15 flowers. **Floral bracts** 1–2 mm long, equalling the pedicels, entire, abaxially scattered lepidote, glabrescent. **Flowers** spreading, suberect, 3–5 mm apart. **Pedicels** 1–2 mm long. **Sepals** 2 mm long, green, glabrous. **Petals** 5 mm long, white, recoiled like watchsprings during anthesis and afterwards. **Filaments** 2 mm long. **Anthers** 2 mm long. **Style** 4 mm long. **Stigmatic complex** simple-erect. **Capsule** not known. **Seeds** not known.

Specimens seen

PERÚ: **Dpto. Puno:** Prov. Carabaya: Between Olleachea and Quillabamba, 13° 30' S, 17°18' W, 1200 m, 12 July 1966, *Vargas 17540* [CUZ!, US (phot!)]; ibid.: Between Olleachea and San Gabán, (13° 30' S, 70° 20' W), 1000–2000 m, 17–24 July, *Dillon et al 1171* [F!].

Distribution and ecology

Range size: Very small.
Countries: PERÚ. Dpto. Cuzco, Puno.
Ecoregions: Peruvian Yungas.
Life style & habitats: Terrestrial or saxicolous in humid and evergreen montane forests.
Altitude: 900–1500 m.

Taxonomic delimitation and systematic relationships

Fosterella aletroides bears some morphological resemblance to *F. batistana* from the Brazilian Amazon rainforest, which is known from the type locality only. But in contrast to *F. batistana*, *F. aletroides* has smaller leaf blades, shorter pedicels and petals recoiled like watchsprings. It also differs in an ecological aspect because *F. aletroides* lives at an altitude of 900 m elevation in the Andes. Hitherto no molecular data is available for *F. aletroides*.

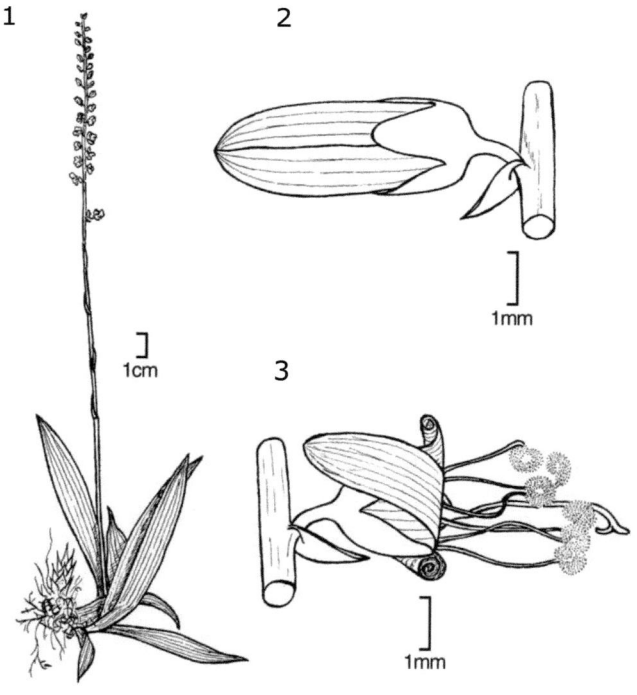

Fig. 11: *Fosterella aletroides*. 1 Habit. 2 Flower before anthesis. 3 Flower during anthesis, one petal uncoiled. Drawing of *Vargas 7358* by R.J. Downs in SMITH (1954). © Smithsonian Institution.

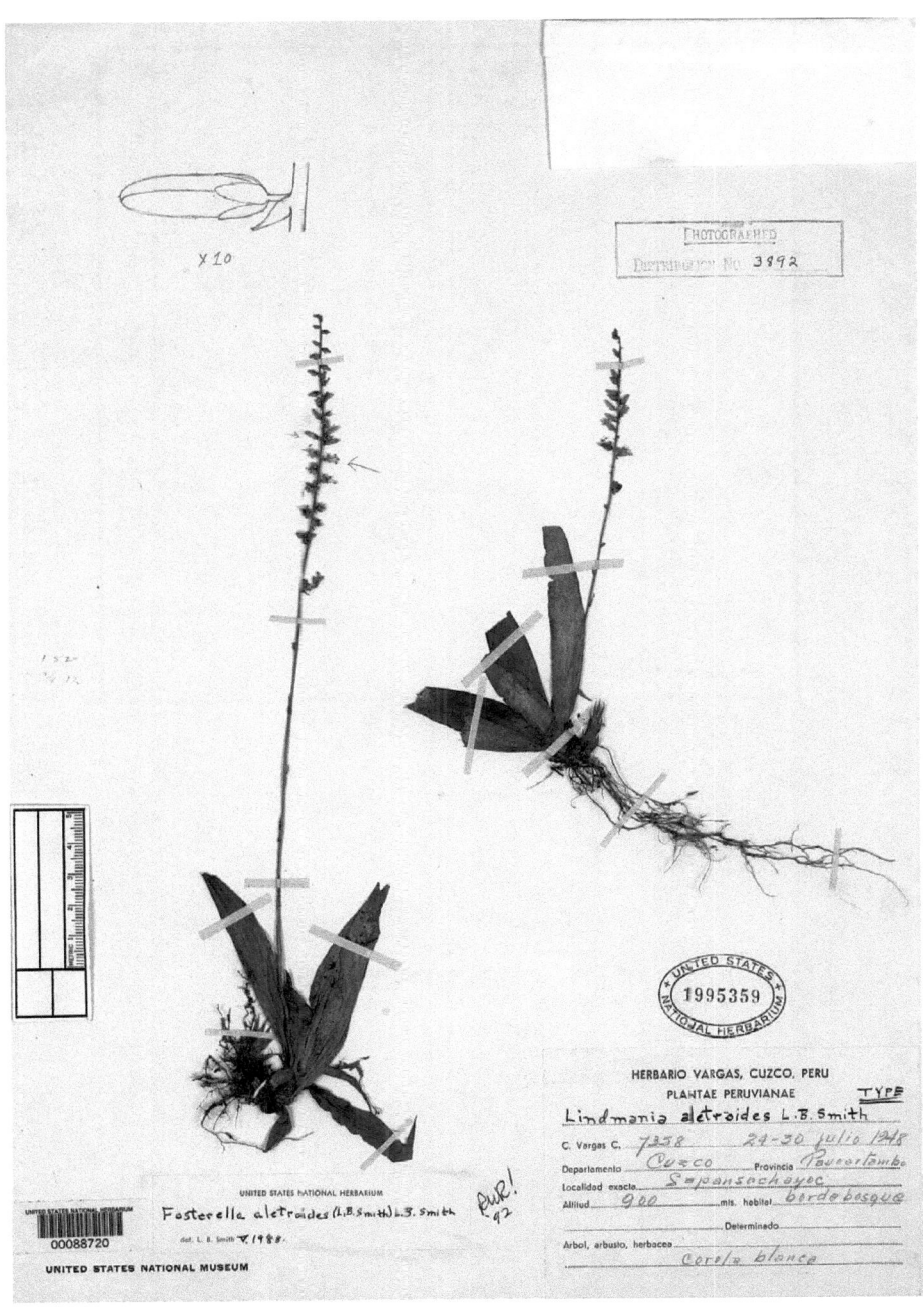

Fig. 12: Holotype of *Fosterella aletroides* (L.B. Sm.) L.B. Sm. [US].

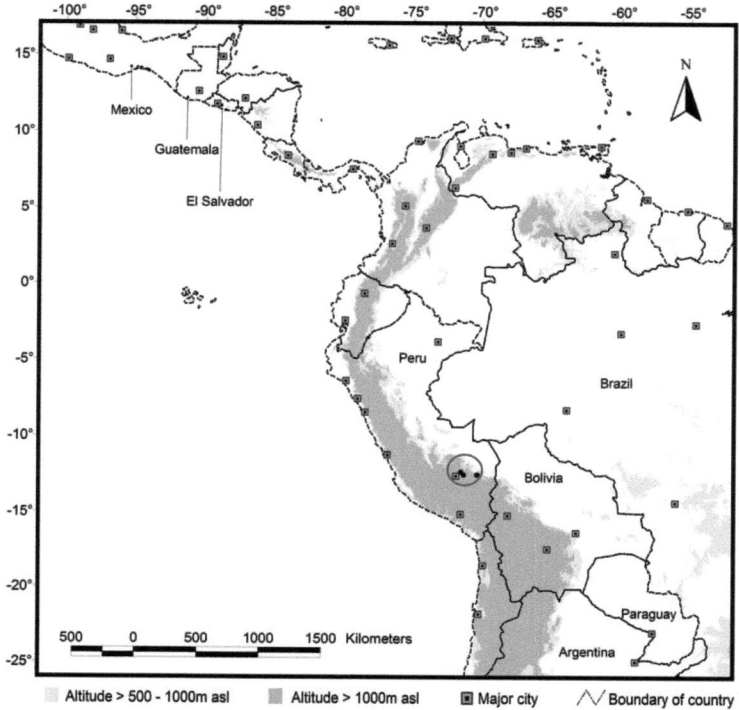

Fig. 13: Distribution of *Fosterella aletroides*.

Fosterella batistana Ibisch, Leme & J. Peters, Selbyana 29 (2): 183. 2008. Fig. 14–18.

    TYPE: Brazil. Estado Pará: Município de Itaituba, 400 km on the road from Santarém to Cuiabá, Curuá-una waterfall (west side of Serrá do Cachimbo), *J.B. Fernandes da Silva s.n.*, cultivated specimen flowered in Jan. 2001, *Leme 5078*. [Holotype: HB!; Isotype: SEL!, LPB!, FR!].

Comment on type

The type locality is situated approx. at 08° 43' S, 55° 02' W. Clonotypes are cultivated in the Botanical Gardens Heidelberg (Germany) and in the private living collection of Elton Leme, Teresópolis (Brazil).

Etymology

This species is dedicated to the collector of the type-specimen, the Amazonian orchid specialist João Batista Fernandes da Silva.

Description

Plant acaulescent, up to 30 cm high. **Leaves** few, up to 10, forming an open, flat rosette. **Sheaths** to 20 mm wide, entire, whitish, glabrous. **Blades** lanceolate, acuminate, narrowed towards the base, to 20 cm long and 3 cm wide, thin, entire, undulate towards the base, adaxially glabrescent, abaxially whitish and scattered lepidote by peltate trichomes with dentate margin, glabrescent. **Peduncle** to 15 cm long, 1–2 mm in diameter, green, arachnoid. **Peduncle bracts** to 1 cm long, shorter than the internodes, entire, abaxially arachnoid. **Inflorescence** racemose or compound racemose, rarely paniculate with short branches of 2nd order, to 15 cm long and 8 cm wide, axes green, sparsely arachnoid. **Primary bracts** up to 10 mm long, shorter than or equalling the sterile base of the branches, entire, abaxially sparsely arachnoid. **Branches** up to 8, inclined, straight, to 7 cm long, bearing up to 15 flowers. **Secondary branches** to 2 cm long, bearing up to 4 flowers. **Floral bracts** 2–3 mm long, equalling the pedicels, entire, abaxially sparsely arachnoid. **Flowers** subsecund, pendulous, 3–5 mm apart. **Pedicels** 2–3 mm long. **Sepals** 2 mm long, green, glabrous. **Petals** 6 mm long, white, recurved during anthesis, straight afterwards. **Filaments** 3–4 mm long. **Anthers** 2 mm long. **Style** 3 mm long. **Stigmatic complex** simple-erect. **Capsule** ovoid, 3 mm long, 2 mm wide. **Seeds** filiform, bicaudate, 2 mm long.

Specimens seen

Hitherto this new species is recorded from the type locality only.

Distribution and ecology

Range size:	Very small.
Country:	BRAZIL. Estado Pará.
Ecoregion:	Madeira-Tapajós-Moist-Forests.
Life style & habitat:	Very humid site within moist, evergreen lowland Amazon rainforest (*Fernandes da Silva s.n.* [HB]).
Altitude:	Not specified.

Fosterella batistana was found as a saxicolous plant at a very humid site, near Curuá-una waterfall, in the middle of the Amazon forest. The collection site is about 400 km from the city of Santarém, far away from any other known localities of *Fosterella*.

Taxonomic delimitation and systematic relationships

Morphologically, *Fosterella batistana* is somewhat close to the Andean species *F. weberbaueri*, but it differs by lepidote leaf blades and arachnoid inflorescence. The separate position of *F. batistana* is underlined by its very special habitat requirements. Molecular data (REX et al. 2009) also indicate a close relationship as well as a clear separation of these two species, which form a well defined and supported group. *Fosterella batistana* also bears some morphological resemblance to *F. aletroides* from the Peruvian Andes. But in contrast to *F. aletroides*, *F. batistana* has larger leaf blades, longer pedicels and only slightly recurved petals.

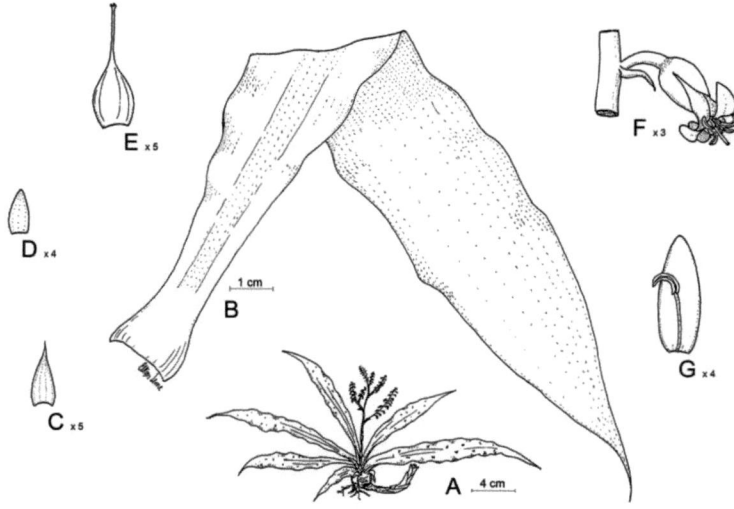

Fig. 14: *Fosterella batistana*. A Habit. B Leaf. C Floral bract. D Sepal. E Gynoecium. F Flower and floral bract. G Petal and stamen. Drawing of *Fernandes da Silva s.n.* by E. Leme in PETERS et al. (2008 b). © Selbyana.

Fig. 15: Holotype of *Fosterella batistana* Ibisch, Leme & J. Peters [HB].

Fig. 16: *F. batistana* (*Fernandes da Silva s.n.*).
Photo: E. Leme.

Fig. 17: Flowers of *F. batistana*
(*Fernandes da Silva s.n.*). Photo: E. Leme.

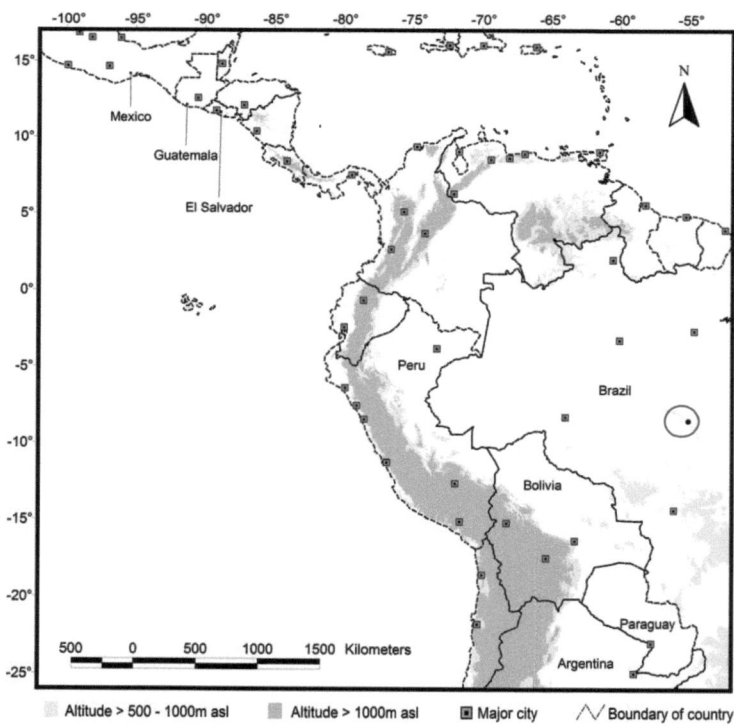

Fig. 18: Distribution of *Fosterella batistana*.

Fosterella caulescens Rauh, Trop. Subtrop. Pflanzenwelt 31: 23. 1979. Fig. 19–23.

TYPE: Bolivia. Between La Paz and Catanavia, degraded lower Yungas, terrestrial, 1200 m, 13 Aug. 1976, *W. Rauh 40579 a* [Holotype: HEID!].

Comment on type

Certainly "Catanavia" means Caranavi, a small town in the Dpto. La Paz: Prov. Nor Yungas, 160 km southeast of the city of La Paz, approx. at 15° 15' S, 67° 34' W, 600 m. "Catanavia" could neither be found on any map, nor did any local people know the name.

Clonotypes are cultivated in the Botanical Garden Berlin, Hamburg, Heidelberg (all Germany), Vienna (Austria), Marie Selby Botanical Garden, Sarasota (Florida) and in the private living collection of Elton Leme, Teresópolis (Brazil).

Etymology

The epithet refers to the species' remarkable caulescence.

Description

Plant caulescent, up to 200 cm high. **Leaves** many, up to 30, forming a dense, more or less upright rosette. **Sheaths** to 50 mm wide, entire, whitish, glabrous. **Blades** narrowly oblanceolate, acuminate, narrowed towards the base, 30–60 cm long and 2–3 cm wide, succulent and serrate towards the base, adaxially sparsely appressed lepidote towards the base, abaxially densely appressed lepidote by peltate trichomes with fimbriate margin. **Peduncle** 50–90 cm long, 5–7 mm in diameter, green, densely arachnoid. **Peduncle bracts** to 10 cm long, longer than the internodes, slightly serrate, abaxially densely appressed arachnoid. **Inflorescence** paniculate with branches up to 2nd and rarely 3rd order, 30–80 cm long and 15–25 cm wide, axes green, arachnoid, glabrescent. **Primary bracts** 15–30 mm long, longer than the sterile base of the branches, entire, abaxially appressed arachnoid. **Branches** up to 25, inclined, arcuate, 10–15 cm long, bearing up to 40 flowers. **Secondary branches** 2–5 (9) cm long, bearing up to 25 flowers. **Floral bracts** 5–7 mm long, longer than the pedicels, entire, abaxially sparsely arachnoid. **Flowers** spreading, suberect, subsessile, 1–5 mm apart. **Pedicels** to 0.5 mm long. **Sepals** 5 mm long, green, sparsely arachnoid. **Petals** 7–8 mm long, pale greenish, recoiled like watchsprings during anthesis and afterwards. **Filaments** 5–7 mm long. **Anthers** 2 mm long. Style 4 mm long. **Stigmatic complex** conduplicate-spiral. **Capsule** ovoid, 4–5 mm long, 2–3 mm wide. **Seeds** filiform, bicaudate, 3 mm long.

Specimens seen

BOLIVIA: **Dpto. La Paz:** Prov. Franz Tamayo: Hills between Río Machariapo and Río Ubito, 14° 25' S, 68° 31' W, 900 m, 22 July 1993, *Kessler 4111* [LPB!, SEL!]; ibid.: Area Natural de Manejo Integrado Madidi, between Azariamas and Sipia, 14° 26' 06" S, 68° 32' 11" W, 920 m, 21 May 2006, *Fuentes & Miranda 10901* [LPB!, MO!]; ibid.: Between Chaquimayo and Tuichi, ca. 20 km NW of Apolo, forest along Río Machariapo, 14° 34' S, 68° 28' W, 1000 m, 12 June 1990, *Gentry 71159* [MO!, SEL!]; ibid.: Río Bilipisa, 10 km NW of Apolo, 14° 36' S, 68° 27' W, 1100 m, 4 July 1997, *Kessler 11000* [LPB!, SEL!]; ibid.: Río Bilipisa, 10 km NW of Apolo, 14° 36' S, 68° 27' W, 1100 m, 4 July 1997, *Kessler 11014* [LPB!, SEL!]. Prov. Nor Yungas: 55 km from Coroico to Caranavi, 16° 12' S, 67° 42' W, 800 m, 1983, *Besse 1775* [SEL!; MSBG]; ibid.: 55 km from Coroico to Caranavi, 16° 12' S, 67° 42' W, 800 m, 1983, *Besse 1776* [SEL!; MSBG]; ibid.: 46 km from Yolosa to Caranavi, between San Pedro and Chojna, 1–2 km before Chojna, 15° 59' 14" S, 67° 35' 23" W, 923 m, 26 Oct. 2002, *Vásquez 4654 b* [VASQ!; FAN!]; ibid.: 2 km from Choro to San Pedro, next to Río Alto Choro, 16° 01' 48" S, 67° 37' 18" W, 800 m, 26 Oct. 2002, *Rex & Schulte 261002-1* [FAN!]. **Dpto. Cochabamba:** Without precise locality, *s.n.* [HEID 601936!].

Distribution and ecology

Range size:	Small.
Countries:	BOLIVIA. Dpto. La Paz, Cochabamba.
Ecoregions:	Sub-Andean-Amazon-Forests, Yungas.
Life style & habitats:	Humid, evergreen lowland and montane forests, terrestrial or saxicolous on steep slopes. Subandean forest with *Anadenanthera colubrina* (Fabaceae) and *Trichilia* sp. (Meliaceae) (*Fuentes & Miranda 10901* [LPB]). Steep slope in humid montane forests, in shady understory with bryophytes and pteridophytes (*Rex & Schulte 261002-1* [FAN]).
Altitude:	800–1200 m.

Taxonomic delimitation and systematic relationships

Morphologically, *Fosterella caulescens* is similar to *F. albicans* (Griseb.) L.B. Sm. but differs by remarkable caulescence, a lax and pendent inflorescence and less pronounced indumentum of the abaxial feaf surface.

According to molecular data (REX et al. 2009), *F. caulescens* is very close to *Fosterella albicans*, *F. kroemeri* and *F. rexiae*.

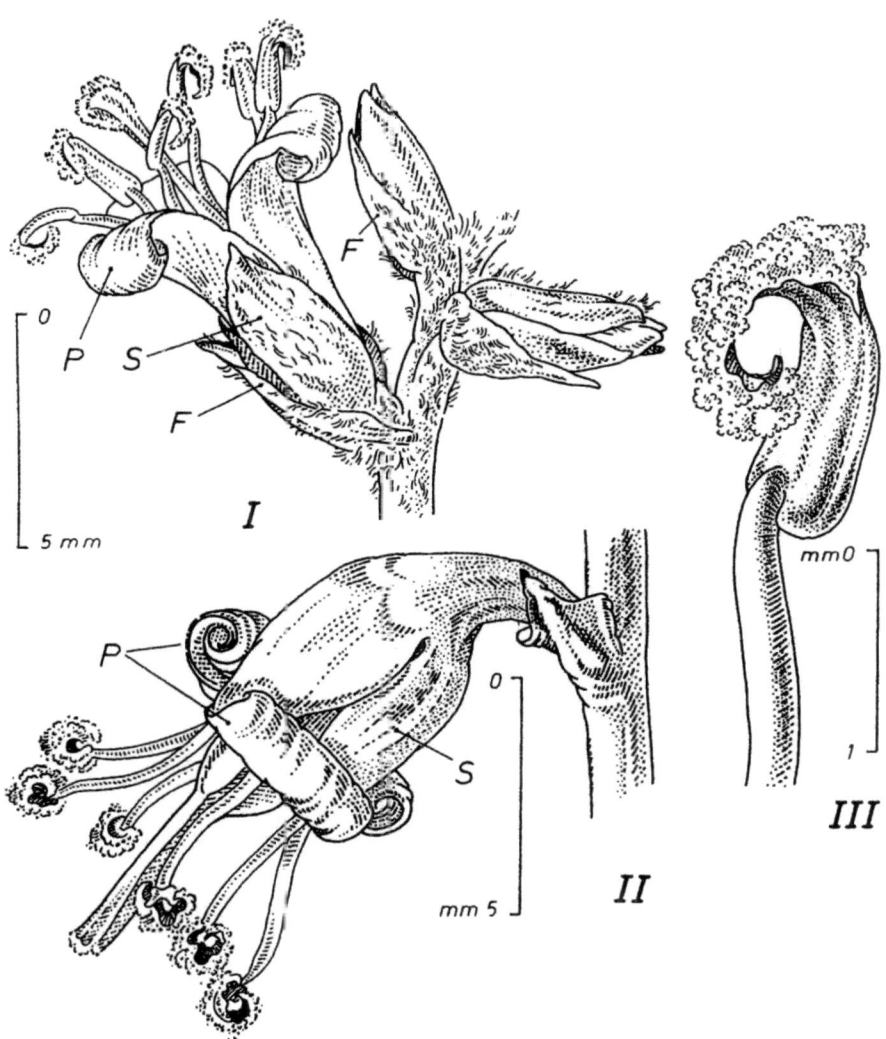

Fig. 19: *Fosterella caulescens*. I Flower during anthesis. II Flower after anthesis. III Stamen. F Floral bract. S Sepal. P Petal. Drawing of *Rauh 40579 a* by F. Rückert in RAUH (1979). © Akademie der Wissenschaften und der Literatur Mainz.

Fig. 20: Holotype of *Fosterella caulescens* Rauh [HEID].

Fig. 21: *F. caulescens (Rauh 40579 a)*.

Fig. 22: Flowers *F. caulescens (Rauh 40579 a)*.

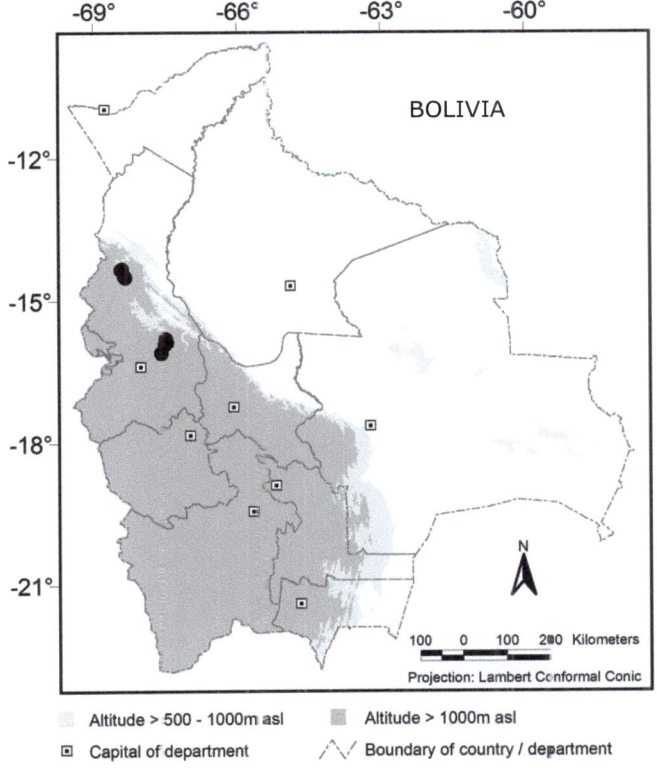

Fig. 23: Distribution of *Fosterella caulescens*.

Fosterella chaparensis Ibisch, R. Vásquez & E. Gross, Rev. Soc. Boliviana Bot., 2 (2): 118. 1999. Fig. 24–28. TYPE: Bolivia. Dpto. Cochabamba: Prov. Chapare, between San Rafael and El Palmar, on humid slope of very humid piedmont rain forest, 65° 30' S, 17° 05' W, 600 m, 10 Aug. 1997, *R. Vásquez & M.I. Vásquez 2792* [Holotype: LPB!; Isotypes: USZ!, FR!].

Etymology

This species is named after the Chapare Province on the Andean foothills, where it was discovered. This region is characterized by the most humid and very species-rich lowland rainforests of Bolivia.

Description

Plant acaulescent, up to 40 cm high. **Leaves** few, 5–8, forming an open, more or less upright rosette. **Sheaths** to 30 mm wide, entire, whitish, glabrous. **Blades** lanceolate, acuminate to acute, narrowed towards the base, almost petiolate, to 25 cm long and 4.5 cm wide, thin, entire, adaxially glabrous, abaxially scattered lepidote by peltate trichomes with dentate margin, glabrescent. **Peduncle** 25–35 cm long, 1–2 mm in diameter, green, glabrous. **Peduncle bracts** 1–3 cm long, shorter than the internodes, entire, abaxially scattered lepidote, glabrescent. **Inflorescence** compound racemose, very rarely paniculate with branches up to 2nd order, 10–35 cm long and 7–12 cm wide, axes green, glabrous. **Primary bracts** 6–10 mm long, longer than or equalling the sterile base of the branches, abaxially scattered lepidote, glabrescent. **Branches** 7–15, inclined, arcuate, 5–10 cm long, bearing up to 25 flowers. **Floral bracts** 1–2 mm long, shorter than the pedicels, entire, abaxially glabrous. **Flowers** spreading, erect, 1–4 mm apart. **Pedicels** 2–4 mm long. **Sepals** 2–3 mm long, green, glabrous. **Petals** 5–7 mm long, white, recurved during anthesis, straight afterwards. Filaments 3 mm long. **Anthers** 2 mm long. **Style** 3–4 mm long. **Stigmatic complex** simple-erect. **Capsule** narrowly ovoid, 4 mm long, 2 mm wide. **Seeds** filiform, bicaudate, 2 mm long.

Specimens seen

BOLIVIA: **Dpto. La Paz:** Prov. Caranavi: Serrania Bella Vista, 42 km from Caranavi towards Sapecho, 15° 40' S, 67° 29' W, 1400 m, 27 Aug. 1997, *Kessler 11489* [LPB!, SEL!]. Prov. Nor Yungas: Valley of Huarinillas, next to Chairo, Estacion Biológica de Tunquini, Sanda below Harnuni de Tunquini, 16° 13' S, 67° 51' W, 1600 m, 2 July 2004, *Beck 30178* [LPB!]; ibid.: Valley of Río Huarinillas towards Chairo, Jucupi, Colgante-Bridge, 16° 12' S, 67° 52' W, 1400 m, 17 Sept. 1995, *Beck 22617* [LPB!, SEL!, USZ!]; ibid.: Cotapata National Park, 16° 12' S, 67° 51' W, 1550 m, 24 Aug. 1999, *Krömer & Acebey 730* [LPB!, SEL!]. Prov. Sud Yungas: Between Unduavi and Chulumani, 2 km from Yanacachi towards Puente Villa, along the Río Unduavi, (16° 24' S, 67° 30' W), 1500 m, 9 Sept. 1989, *Beck 16883* [LPB!, SEL!]. **Dpto. Beni:** Prov. Ballivian: 12 km from Yucumo towards Rurrenabaque, 15° 04' S, 67° 07' W, 450 m, 24 July 1997, *Kessler 10797* [SEL! (st.)]. **Dpto. Cochabamba:** Prov. Chapare: Road from Cochabamba to Villa Tunari, 17° 07' 12" S, 65° 35' 00" W, 1500 m, 28 June 1996, *Kessler 7881* [LPB!]; ibid.: 143 km on the old road between Cochabamba and Villa Tunari, 17° 07' S, 65° 34' W, 1300 m, 25 Aug. 1996, *Kessler 7786* [B!, FR!, LPB!, SEL!, WU!; BGB!, BGGÖ!].

Distribution and ecology

Range size:	Small.
Countries:	BOLIVIA. Dpto. La Paz, Beni, Cochabamba.
Ecoregions:	Sub-Andean-Amazon-Forests, Yungas.
Life style & habitats:	In understory of more or less disturbed humid, evergreen lowland and montane forests. Terrestrial or saxicolous, on shady, humid slopes along roadsides.
Altitude:	450–1600 m.

Taxonomic delimitation and systematic relationships

This taxon differs from all *Fosterella* species with slightly recurved or straight petals and glabrous leaves by the erect and spreading flowers. It is quite similar to *F. weberbaueri*, but differs in its more or less upright rosette and and only slightly recurved petals. Up to now, no molecular data is available for *F. chaparensis*.

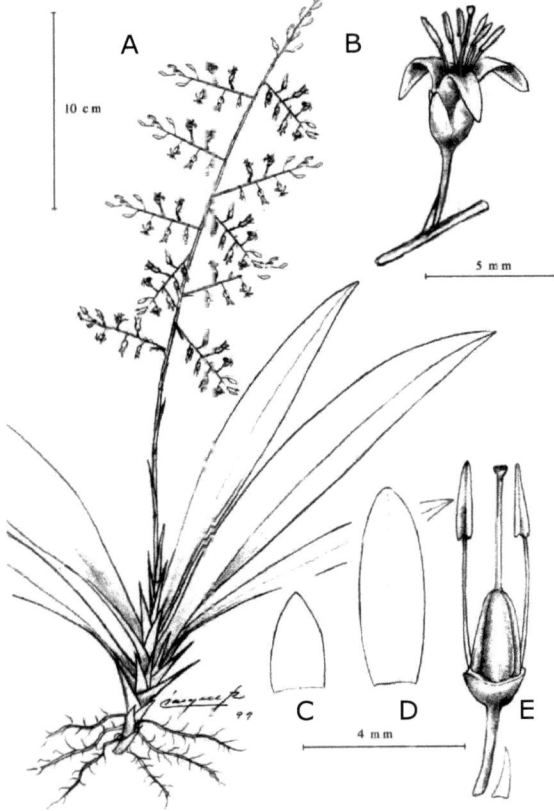

Fig. 24: *Fosterella chaparensis*. A Habit. B Flower. C Sepal. D Petal. E Gynoecium and androecium. Drawing of *Vásquez & Vásquez 2792 a* by R. Vásquez in IBISCH et al. (1999). © Sociedad Boliviana de Botánica.

Fig. 25: Holotype of *Fosterella chaparensis* Ibisch, R. Vásquez & E. Gross [LPB].

Fig. 26: *F. chaparensis (Kessler 7786)*.

Fig. 27: Flowers of *F. chaparensis (Kessler 7786)*.

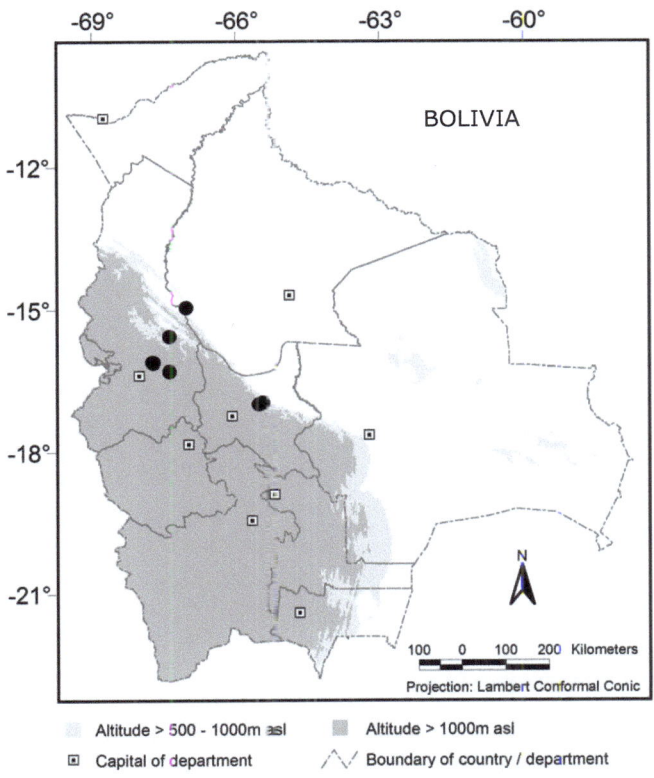

Fig. 28: Distribution of *Fosterella chaparensis*.

Fosterella christophii Ibisch, R. Vásquez & J. Peters, Selbyana 29 (2): 185. 2008. Fig. 29–32.

TYPE: Bolivia. Dpto. Santa Cruz: Prov. A. Ibáñez, Municipio de Terebinto, Arubay, 8 km north of Terebinto, nearby Rio Guendá and the property of G. Coimbra, in understory of semihumid forest, 17° 44' S, 63° 23' W, 450 m, 16 Nov. 1998, *P. Ibisch & C. Nowicki 98.0173* [Holotype: LPB!; Isotypes: FR!, SEL!, USZ!, WU!].

Comment on type

Clonotypes are cultivated in the Botanical Gardens Heidelberg (Germany) and in the living collection of the Fundación Amigos de la Naturaleza, Santa Cruz (Bolivia).

Etymology

The taxon is dedicated to the German biologist Christoph Nowicki (*1969), who permanently supported the *Fosterella* research and was co-collector of the type specimen.

Description

Plant acaulescent, up to 100 cm high. **Leaves** many, 25–30, forming a dense, arched rosette. **Sheaths** to 15 mm wide, entire, whitish, glabrous. **Blades** broadly lanceolate to oblanceolate, acuminate to cuspidate, 20–50 cm long and 3–7 cm wide, thin, entire, adaxially glabrous, abaxially frequently reddish, scattered lepidote by peltate trichomes with dentate margin. **Peduncle** 30–60 (80) cm long, 3–5 (7) mm in diameter, frequently reddish, densely villous-arachnoid. **Peduncle bracts** 3–5 cm long, equalling the internodes, entire, frequently reddish, abaxially scattered lepidote. **Inflorescence** compound racemose or paniculate with lateral branches up to 2nd order, 10–40 cm long and 5–15 cm wide, axes frequently reddish, villous. **Primary bracts** 10–20 mm long, equalling the sterile base of the primary branches, entire, frequently reddish, abaxially villous. **Branches** 10–15, ascending, straight, 5–10 cm long, bearing up to 30 flowers. **Secondary branches** to 7 cm long, bearing up to 15 flowers. **Floral bracts** 3–5 mm long, always longer than the pedicels, entire, abaxially sparsely villous. **Flowers** secund, pendulous, 3–4 mm apart. **Pedicels** 2–3 mm long. **Sepals** 2–3 mm long, frequently reddish, sparsely villous. **Petals** 6–7 mm long, white, recurved during anthesis, straight afterwards. **Filaments** 4 mm long. **Anthers** 2 mm long. **Style** 3 mm long. **Stigmatic complex** conduplicate-spiral. **Capsule** ovoid, 4–5 mm long, 2 mm wide. **Seeds** filiform, bicaudate, 2 mm long.

Specimens seen

BOLIVIA: Without precise locality, July 1996, *Pardo s.n.* [HB!; LEME!]. **Dpto. Santa Cruz:** Prov. Florida: 9 km SE of Bermejo, Río Sillar, old road from Santa Cruz to Cochabamba, 18° 10' S, 63° 35' W, 800 m, 25 Sept. 1990, *Vargas 738* [LPB!, MO!, NY!, SEL!, USZ!]; ibid.: 5 km above Bermejo, 900 m, 18 Sept. 1996, *Wood 11343* [K!]; ibid: Between Angostura and Samaipata, 18° 06' S, 63° 20' W, 11 June 2001, *Mendoza 110* [LPB!, NY!, USZ!]; ibid.: Laguna Volcanes, 1100 m, 2002, *Ibisch 02.0002* [FR!, HEID!, LPB!, SEL!, USZ!, WU!; FAN!, BGHD!]; ibid.: Next to Samaipata,

(18° 12' S, 63° 48' W), 1500 m, 22 Mar. 1994, *Aichinger s.n.* [M!]. <u>Prov. Andres Ibañez:</u> 3 km from La Guardia to Sargento del Torno, Nov. 1999, *Justiniano s.n.* [LPB , SEL!]; ibid.: Arubay, 8 km N of Terevinto, property of G. Coimbra, nearby Río Guendá, 17° 41' S, 63° 25' W, 460 m, 4 Oct. 1998, *Vásquez 2994* [FR!, LPB!, VASQ!; FAN!]; ibid.: Between Santa Cruz and Samaipata, *Correale s.n.* [F!; SEL!; MSBG].

Distribution and ecology

Range size:	Very small.
Countries:	BOLIVIA. Dpto. Santa Cruz.
Ecoregions:	Pre- and Sub-Andean-Amazon-Forests in contact to Chiquitano-Dry-Forest.
Life style & habitats:	Semihumid and semideciduous lowland forests at the transition from southwest Amazon to Chiquitano-Dry-Forest and at the transition from Yungas to Tucuman-Bolivian-Forest and Inter-Andean-Dry-Forests (Andean knee). Terrestrial or saxicolous in understory of more or less closed forests, on sandy ground and shady, humid slopes.
Altitude:	450–1500 m.

Taxonomic delimitation and systematic relationships

Morphologically, *Fosterella christophii* is very similar to *F. villosula*, but differs in the slightly less villous inflorescence with shorter floral bracts, sepals and petals. Another important difference is its ecology, as *F. villosula* mainly occurs in the very humid rain forest of the Chapare region, Cochabamba, whereas *F. christophii* has a very small distribution range at the Andean knee, close to the city of Santa Cruz. Molecular findings (REX et al. 2009) strongly support the close relationship of these two species as well as their clear separation. Together with *F. micrantha* from Central America they form a very clearly delimited and highly supported group.

Fig. 29: Holotype of *Fosterella christophii* Ibisch, R. Vásquez & J. Peters [LPB].

Fig. 30: *F. christophii* (Ibisch & Nowicki 98.0173).

Fig. 31: Flowers of *F. christophii* (Ibisch & Nowicki 98.0173).

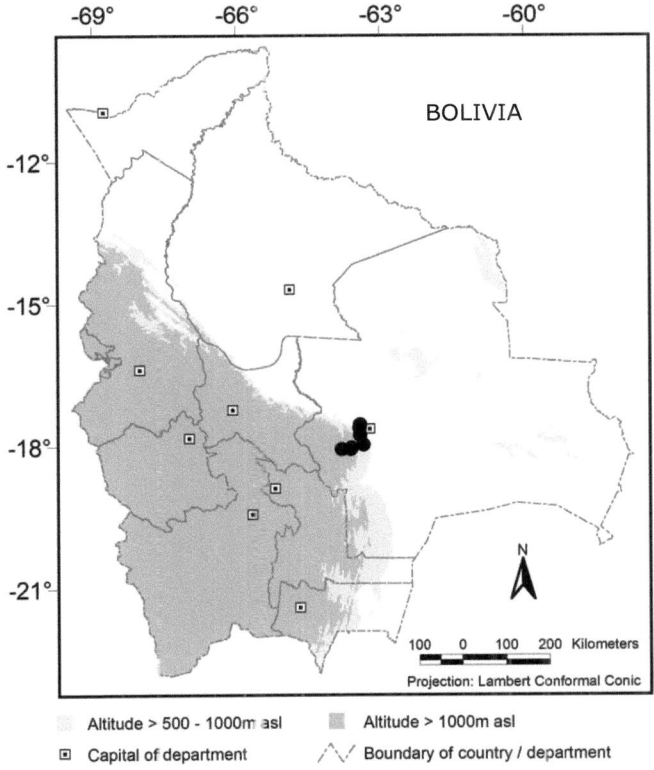

Fig. 32: Distribution of *Fosterella christophii*.

Fosterella cotacajensis M. Kessler, Ibisch & E. Gross, Rev. Soc. Boliviana Bot. 2 (2): 111. 1999. Fig. 33–37. TYPE: Bolivia. Dpto. Cochabamba: Prov. Ayopaya, 21 km from Cocapata to Cotacajes, 16° 46' S, 66° 44' W, 2100 m, 16 May 1997, *M. Kessler, J. Gonzales, K. Bach & A. Acebey 9620* [Holotype: LPB!; Isotype: SEL! (st.)].

Etymology

The epithet refers to the valley of Río Cotacajes, to which the distribution of this species probably is limited.

Description

Plant caulescent, up to 100 cm high. **Leaves** many, 15–25, forming a dense, more or less upright rosette. **Sheaths** to 30 mm wide, entire, greenish, glabrous. **Blades** lanceolate, 15–45 cm long and to 2 cm wide, succulent and serrate towards the base, adaxially sparsely lepidote, abaxially covered by a thick layer of interwoven, peltate trichomes with fimbriate margin. **Peduncle** 20–50 cm long, 5 mm in diameter, glabrous, glaucous. **Peduncle bracts** 3–5 cm long, much longer than the internodes, slightly serrate, abaxially by a layer of interwoven trichomes. **Inflorescence** paniculate with lateral branches up to 2nd order, to 30 cm long and 10 cm wide, axes green, glabrous, glaucous. **Primary bracts** 10–30 mm long, longer than or equalling the sterile base of the branches, entire, abaxially covered by interwoven trichomes. **Branches** 10–15, ascending, straight, 5–15 cm long, bearing up to 25 flowers. **Secondary branches** 2–3 cm long, bearing up to 5 flowers. **Floral bracts** 4–7 mm long, always longer than the pedicels, entire, abaxially glabrous. **Flowers** secund, pendulous, 5–10 mm apart. **Pedicels** 3–4 mm long. **Sepals** 5 mm long, green, glabrous. **Petals** 8 mm long, white, recoiled like watchsprings during anthesis and afterwards. **Filaments** 3 mm long. **Anthers** 2 mm long. **Style** 2 mm long. **Stigmatic complex** simple-erect. **Capsule** ovoid, 5–6 mm long, 3–4 mm wide. **Seeds** filiform, bicaudate, 2 mm long.

Specimens seen

BOLIVIA: **Dpto. La Paz:** Prov. Inquisivi: 2 km from Río Miguillas towards Río La Paz, 16° 32' S, 67° 21' W, 1300 m, 24 Sept. 1995, *Kessler 5616* [SEL!]; ibid.: 6 km from Inquisivi to Sita, 16° 53' S, 67° 08' W, 2200 m, 17 Sept. 1995, *Kessler 5531* [LPB!, SEL!]; ibid.: 12 km from Miguillas to La Pazuela, 16° 33' S, 67° 24' W, 1250 m, 22 Sept. 1995, *Kessler 5585* [SEL!]; ibid.: Along the road in Cajuata, 16° 42' S, 67° 10' W, 1800 m, 28 Dec. 1989, *Lewis 36936* [LPB!, MO!, SEL!]; ibid.: Parque Nacional de Choquecamiri, Lakachaka, trail between the mouth of Río Aguilani and Choquecamiri, 16° 40' S, 67° 20' W, 1700 m, 23 Sept. 1991, *Lewis 40425* [LPB!, MO!, SEL!]. **Dpto. Cochabamba:** Prov. Ayopaya: Between Pujiuni and Cotacajes, 16° 44' S, 66° 44' W, 1370 m, 20 June 2001, *Vásquez 4127* [VASQ!]; ibid.: Between Cotacajes and Atispaya, 16° 40' S, 66° 44' W, 1300 m, 21 June 2001, *Vásquez 4815* [VASQ!]; ibid.: 3 km uphill from Sequerancho, road towards Atispaya, 16° 31' 02" S, 66° 46' 20" W, 1550 m, 24 July 2001, *Vargas 6377* [FAN!].

Distribution and ecology

Range size: Very small.

Countries: BOLIVIA. Dpto. La Paz, Cochabamba.

Ecoregion: Inter-Andean-Dry-Forests.

Life style & habitats: Dry deciduous forests in high valleys (probably below 1000 mm mean annual precipitation). Along roadsides and riverbanks, on sunny, dry, rocky slopes in understory of open, semideciduous forests. Terrestrial in deciduous and semideciduous forests dominated by *Acacia macracantha* (Fabaceae) and *Prosopis* sp. (Fabaceae), perturbed localities with *Schinopsis* cf. *haenkeana* (Anacardiaceae), *Aspidosperma quebracho-blanco* (Apocynaceae) and *Jacaranda mimosifolia* (Bignoniaceae). Due to frequent sheep-run, the understory is characterized by graze-weeds like *Croton* sp. (Euphorbiaceae), *Puya* sp. (Bromeliaceae) and various Cactaceae (*Pereskia weberiana, Samaipaticereus inquisivensis, Opuntia* sp. and *Cereus huilunchu*) (IBISCH et al. 1999).

Altitude: 1150–2200 m.

Taxonomic delimitation and systematic relationships

According to molecular data (REX et al. 2007), *Fosterella cotacajensis* is very close to *F. weddelliana*, which morphologically is very much alike. But compared to *F. weddelliana*, *F. cotacajensis* has narrower leaf blades, which are not narrowed towards the base and longer floral bracts and sepals. Molecular data also indicate a clear separation of these two species, which are forming a well defined and supported group. Presumably, *F. cotacajensis* has been derived from a *F. weddelliana* ancestor, conquering higher altitudes and more arid habitats.

Fosterella cotacajensis differs from *F. rusbyi* – to which it is quite similar as well – in narrower leaf blades, longer floral bracts and sepals and serrate peduncle bracts.

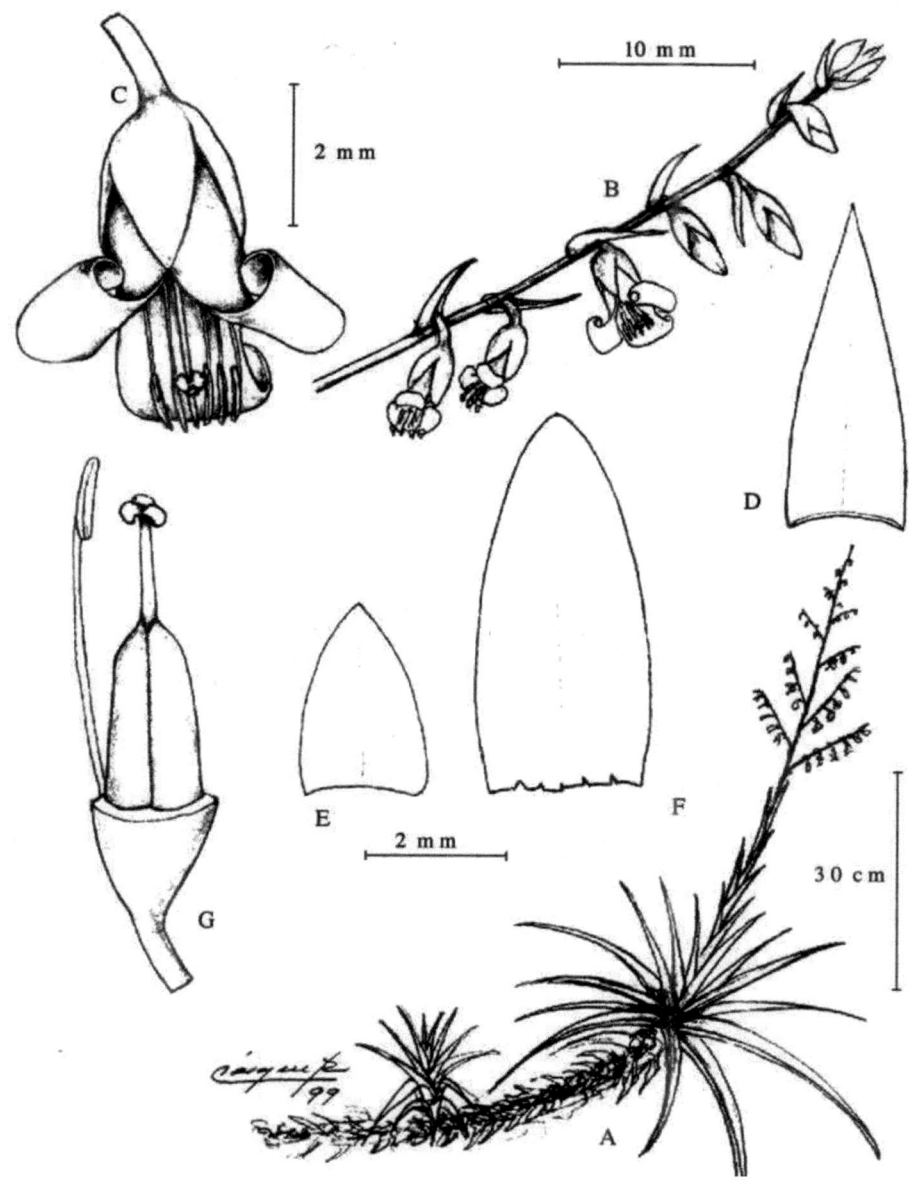

Fig. 33: *Fosterella cotacajensis*. A Habit. B Floral branch. C Flower. D Floral bract. E Sepal. F Petal. G Gynoecium and androecium. Drawing of *Kessler 9620* by R. Vásquez in KESSLER et al. (1999). © Sociedad Boliviana de Botánica.

Fig. 34: Holotype of *Fosterella cotacajensis* M. Kessler, Ibisch & E. Gross [LPB].

Fig. 35: *F. cotacajensis (Kessler 9620).*

Fig. 36: Flowers of *F. cotacajensis (Vásquez 3620).*
Photo: P. Ibisch.

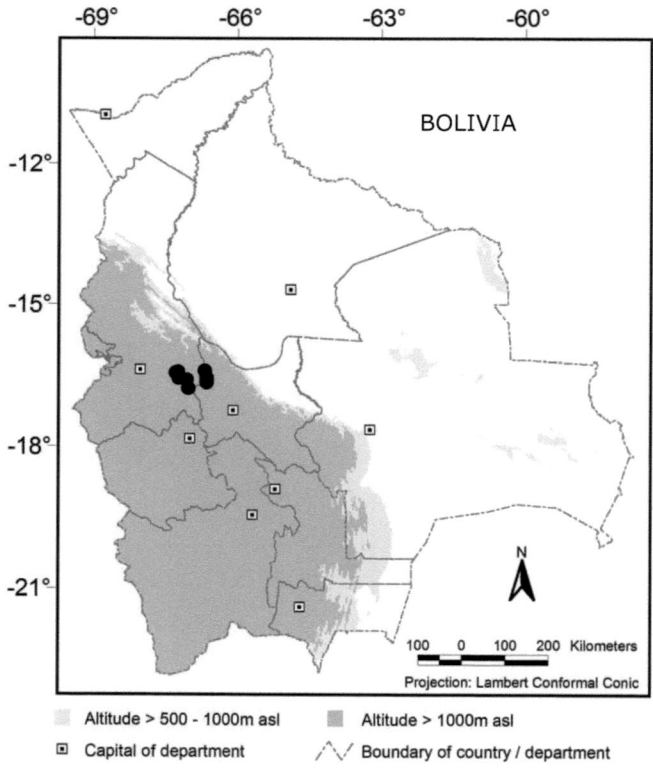

Fig. 37: Distribution of *Fosterella cotacajensis.*

Fosterella elviragrossiae Ibisch, R. Vásquez & J. Peters, Selbyana 29 (2): 188. 2008. Fig. 38–41.
TYPE: Bolivia. Dpto. Cochabamba: Prov. Ayopaya, Municipio de Atispaya, 16° 36' S, 66° 43' W, 1330 m. 22 June 2001, *R. Vásquez, G. Navarro, M. Fernández, F. Miranda & H. Rocha 4177 b* [Holotype: LPB!; Isotype: FR!, SEL!].

Etymology
This species is dedicated to the German bromeliad taxonomist Elvira Groß (1954–2005).

Description
Plant acaulescent, up to 35 cm high. **Leaves** few, up to 15, forming an open, flat rosette. **Sheaths** to 20 mm wide, entire, whitish, glabrous. **Blades** lanceolate, narrowed towards the base, to 20 cm long and 2 cm wide, succulent towards the base, entire, undulate towards the base, adaxially lepidote towards the base, abaxially sometimes reddish and densely appressed lepidote by peltate trichomes with fimbriate margin. **Peduncle** to 30 cm long, 1–2 mm in diameter, green or slightly reddish, glabrous, glaucous. **Peduncle bracts** to 1.5 cm long, much shorter than internodes, entire, reddish, abaxially sparsely lepidote, glabrescent. **Inflorescence** racemose or compound racemose, to 20 cm long and 7 cm wide, axes green, glabrous, glaucous. **Primary bracts** to 8 mm long, slightly longer than the sterile base of the primary branches, entire, abaxially sparsely lepidote, glabrescent. **Branches** few, up to 3, ascending, straight, to 4 cm long, bearing up to 8 flowers. **Floral bracts** 2 mm long, longer than or equalling the pedicels, entire, abaxially glabrous. **Flowers** secund, pendulous, 5–15 mm apart. **Pedicels** to 2 mm long. **Sepals** 1.5 mm long, green, glabrous. **Petals** 4–5 mm long, white, recurved during anthesis, straight afterwards. **Filaments** 4 mm long. **Anthers** 1 mm long. **Style** 2 mm long. **Stigmatic complex** simple-erect. **Capsule** narrowly ovoid, 4 mm long, 2 mm wide. **Seeds** filiform, bicaudate, 2 mm long.

Specimens seen
BOLIVIA: **Dpto. Cochabamba:** Prov. Ayopaya: 16° 31' 02" S, 66° 46' 20" W, 1150 m 24 July 2001, *Vargas 6387* [FR!, LPB!, SEL!; FAN!].

Distribution and ecology

Range size:	Very small
Country:	BOLIVIA. Dpto. Cochabamba.
Ecoregions:	Sub-Andean-Amazon-Forests, Yungas.
Ecology & habitats:	Semihumid to humid, evergreen lowland and montane forests. In crevices of dry slopes in open forest.
Altitude:	1150–1330 m.

Taxonomic delimitation and systematic relationships

Fosterella elviragrossiae resembles *F. rojasii* regarding some characters, but differs by smaller stature, less branched inflorescence, leaf blades narrowed towards the base, glabrous peduncle, small scape bracts which are considerably shorter than the internodes, and petals that are only recurved at anthesis and straight afterwards. *Fosterella elviragrossiae* occurs in humid montane forests of the Cochabamba Department while *F. rojasii* is known from the Paraguayan Chaco. Up to now, no molecular data is available for *F. elviragrossiae*.

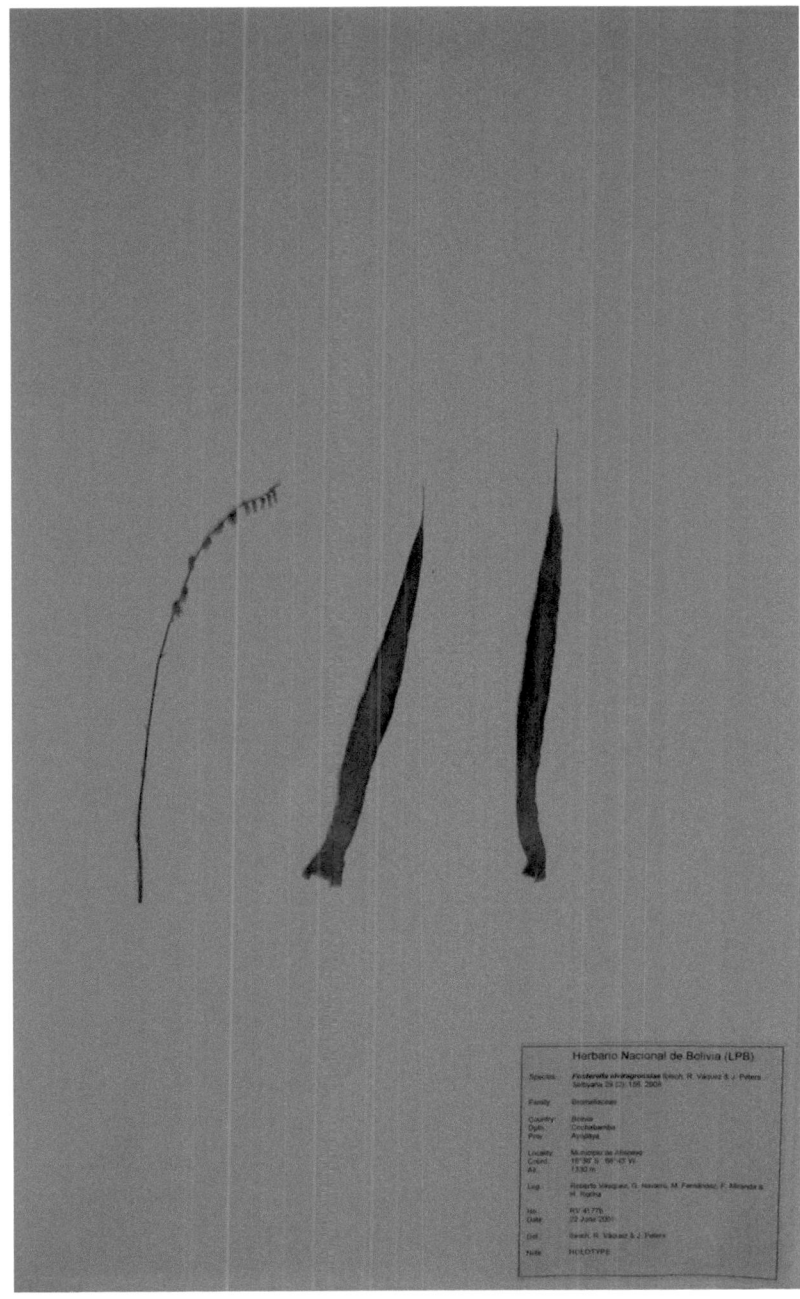

Fig. 38: Holotype of *Fosterella elviracrossiae* Ibisch, R. Vásquez & J. Peters [LPB].

Fig. 39: *F. elviragrossiae (Vásquez 4177 b)*.
Photo: P. Ibisch.

Fig. 40: Flowers of *F. elviragrossiae*
(Vásquez 4177 b). Photo: P. Ibisch.

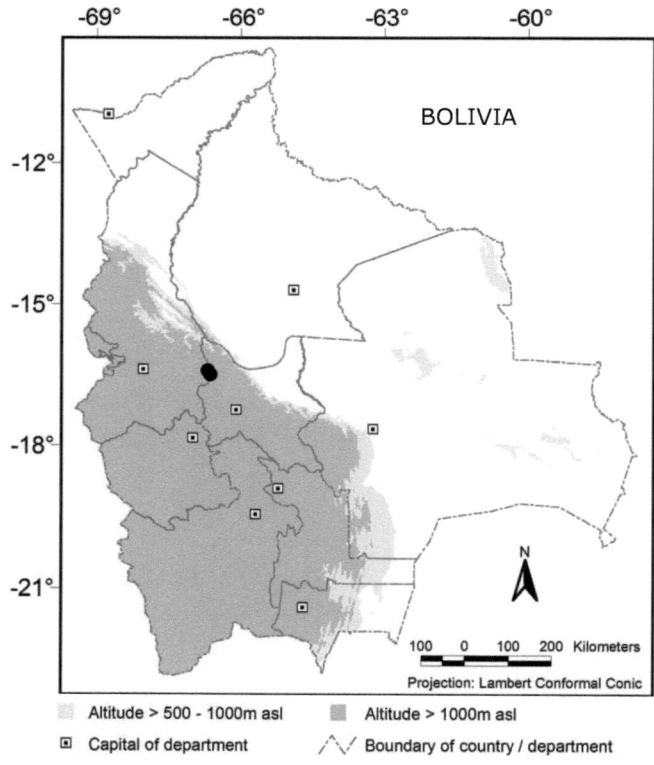
Fig. 41: Distribution of *Fosterella elviragrossiae*.

Fosterella floridensis Ibisch, R. Vásquez & E. Gross, Rev. Soc. Boliviana Bot. 2 (2): 120. 1999. Fig. 42–46. TYPE: Bolivia. Dpto. Santa Cruz: Prov. Florida, Refugio Volcanes, in understory of seminumid forest, 18° 05' S, 63° 40' W, 1100 m, 30 Dec. 1997, P. & C. Ibisch 97.0083, flowered in the garden of Pierre L. Ibisch, Santa Cruz de la Sierra, Jan. 1999 [Holotype: LPB!; Isotype: FR!].

Etymology

The epithet refers to the Province of Florida, Department of Santa Cruz, an area rich in locally endemic Bromeliaceae.

Description

Plant acaulescent, up to 150 cm high. **Leaves** many, 20–25, forming an open, more or less upright rosette. **Sheaths** to 35 mm wide, entire, whitish, glabrous. **Blades** lanceolate acuminate, narrowed towards the base, but not petiolate, 50–70 cm long and 1.5–3 cm wide, succulent towards the base, entire, adaxially glabrous, abaxially densely appressed lepidote by peltate trichomes with fimbriate margin. **Peduncle** 60–90 cm long, to 10 mm in diameter, green, arachnoid. **Peduncle bracts** 5–15 cm long, longer than the internodes, entire, abaxially densely appressed lepidote. **Inflorescence** paniculate with branches up to 2nd order, 25–60 cm long and 7–15 cm wide, axes green, sparsely arachnoid. **Primary bracts** 10–20 mm long, equalling the sterile base of the branches, abaxially appressed lepidote. **Branches** 10–20, ascending, straight, 5–15 cm long, bearing up to 80 flowers. **Secondary branches** 2–5 cm long, bearing 5–10 flowers. **Floral bracts** 2–3 mm long, longer than the pedicels, abaxially sparsely arachnoid. **Flowers** spreading, erect, 1–2 mm apart. **Pedicels** 1–2 mm long. **Sepals** 6–8 mm long, green, sparsely villous-arachnoid. **Petals** 8–12 mm long, white, straight during anthesis and afterwards. **Filaments** 5 mm long. **Anthers** 1 mm long. **Style** 3 mm long. **Stigmatic complex** simple-erect. **Capsule** ovoid, 3 mm long, 2 mm wide. **Seeds** filiform, bicaudate, 2 mm long.

Specimens seen

BOLIVIA: **Dpto. Santa Cruz:** Prov. Ichilo: Without precise locality, 17° 44' S, 63° 37' W, 450 m, 16 Sept. 1996, *Kessler 8612* [LPB!]. Prov. Florida: El Sillar, 1350 m, 5 Nov. 2000, *Vásquez 3796 b* [SEL!; FAN!]; ibid.: Refugio Los Volcanes, 3 km NE of Bermejo, 18° 06' S, 63° 36' W, 1050 m, 3 Oct. 1997, *Kessler 12272* [LPB!, SEL!]; ibid.: 4.5 km from Angostura to Samaipata, 18° 10' S, 63° 35 W, 600 m, 9 Feb. 2002, *Ibisch 02.0001* [SEL!; FAN!]; ibid.: Along the road between Samaipata and Bermejo, 18° 08' 24" S, 63° 41' 09" W, 1019 m, 11 Oct. 2002, *Rex & Schulte 111002-11* [FR!; FAN!]; ibid.: Along the road between Samaipata and Bermejo, 18° 08' 24" S, 63° 41' 09" W, 1019 m, 11 Oct. 2002, *Rex & Schulte 111002-12* [LPB!; FAN!]; ibid.: Along the road between Samaipata and Bermejo, 18° 08' 24" S, 63° 41' 09" W, 1019 m, 11 Oct. 2002, *Rex & Schulte 111002-13* [SEL!; FAN!]; ibid.: Along the road between Samaipata and Bermejo, 18° 08' 24" S, 63° 41' 09" W, 1019 m, 11 Oct. 2002, *Rex & Schulte 111002-14* [FR!; FAN!]; ibid.: Along the road between Samaipata and Bermejo, 18° 08' 24" S, 63° 41' 09" W, 1019 m, 11 Oct. 2002, *Rex & Schulte 111002-15* [SEL!; ibid.: Along the road between Samaipata and Bermejo, 18° 08' 24" S, 63° 41' 09" W, 1019m, 11 Oct. 2002, *Rex & Schulte 111002-16* [LPB!; FAN!].

WITHOUT LOCALITY: 1 Mar. 1984, *Pascal* [SEL!].

Distribution and ecology

Range size: Very small.
Country: BOLIVIA. Dpto. Santa Cruz.
Ecoregions: Sub-Andean-Amazon-Forests at the transition from Yungas to Tucuman-Bolivian-Forest and Inter-Andean-Dry-Forests (Andean knee).
Life style & habitats: Understory of slightly disturbed, semihumid, semideciduous lowland and montane forests. On steep, shady slopes along roadsides.
Altitude: 450–1350 m.

Taxonomic delimitation and systematic relationships

Morphologically, *F. floridensis* differs from all species of *Fosterella* by straight, not recurved petals, the erect (neither secund nor pendulous) flowers and by abaxially lepidote leaf blades. The only other *Fosterella* species with straight petals is *F. spectabilis*, which has glabrescent leaf blades and secund, pendulous, relatively large and red flowers.

In the phylogenetic tree deduced from AFLP-analysis (REX et al. 2009), *F. floridensis* forms a monophylum together with *F. vasqeuzii* and *F. rusbyi*, from which it is morphologically and ecologically clearly distinct. In the phylogeny deduced from DNA sequence data at four chloroplast loci (REX et al. 2009), *F. floridensis* belongs to the "*rusbyi*-group", a clade comprising a set of morphologically distinct species (*F. rusbyi*, *F. vasquezii*, *F. windischii*, *F. yuvinkae* and *F. spectabilis*) from rather diverse environments.

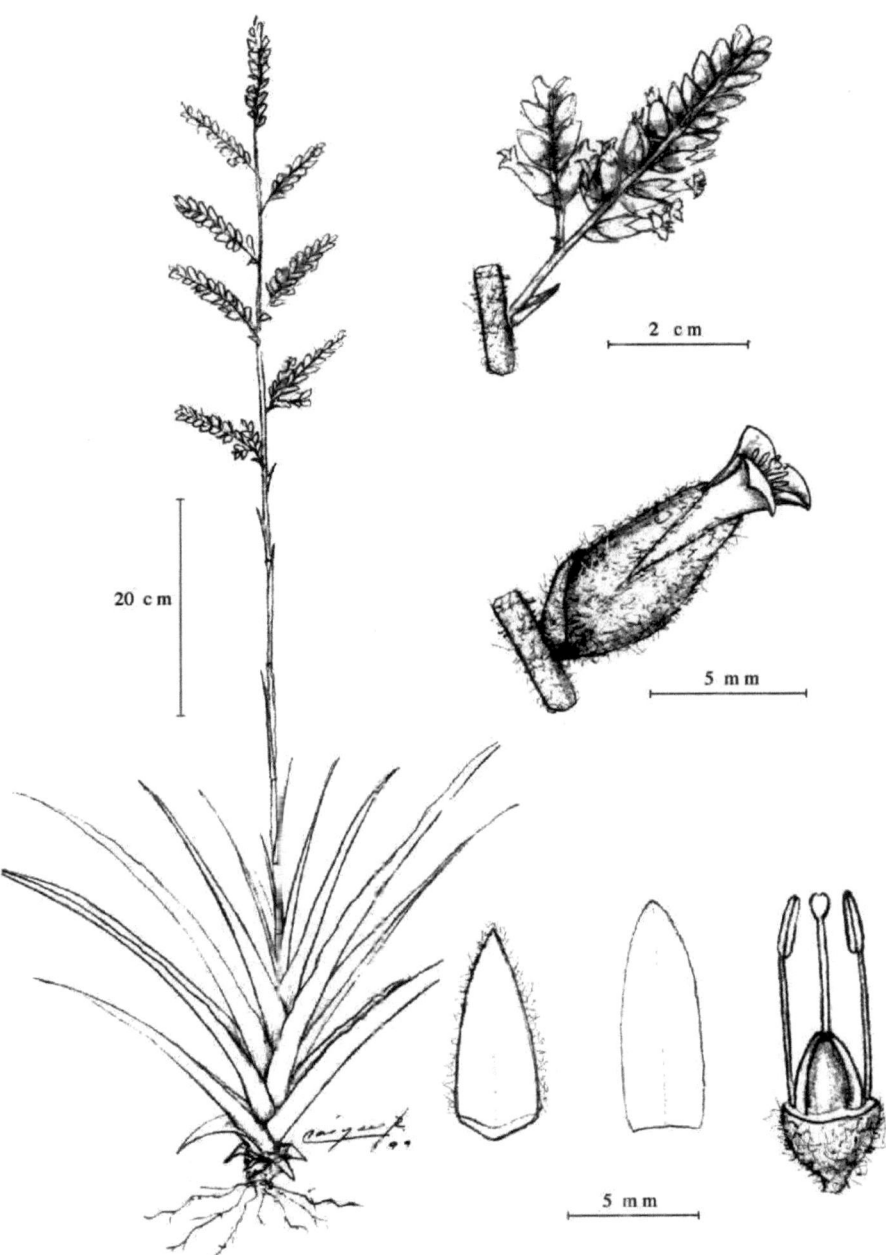

Fig. 42: *Fosterella floridensis*. A Habit. B Floral branch. C Flower. D Sepal. E Petal. F Gynoecium and androecium. Drawing of *Vásquez & Vásquez 2792 a* by R. Vásquez in IBISCH et al. (1999). © Sociedad Boliviana de Botánica.

Fig. 43: Holotype of *Fosterella floridensis* Ibisch, R. Vásquez & E. Gross [LPB].

Fig. 44: *F. floridensis (Ibisch 02.0001)* in the natural habitat. Photo: P. Ibisch.

Fig. 45: Flowers of *F. floridensis (Rex & Schulte 111002-13)*. Photo: P. Ibisch.

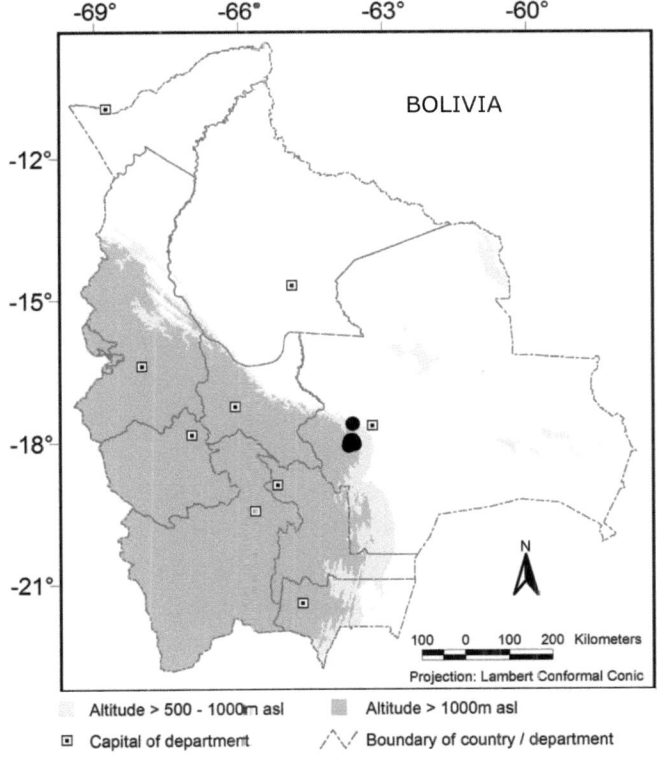

Fig. 46: Distribution of *Fosterella floridensis*

Fosterella gracilis (Rusby) L.B. Sm., Phytologia 7: 171. 1960. Fig. 47–50.

Basionym:	*Catopsis gracilis* Rusby, Bull. New York Bot. Gard. 6: 489. 1910. TYPE: Guanai, 500 m, 27 Sept. 1901, *R.S. Williams 738*. [Lectotype: NY!; Isotype: BM (phot.!), K (phot.!)].
≡	*Lindmania gracilis* (Rusby) L.B. Sm., Contr. Gray Herb. 104: 78. 1934.

Comment on type

SMITH (1934 b) located the type locality Guanai in Bolivia, near Santa Ana, Bopi. This indication can be specified, as Guanay (approx. at 15° 30' S, 67° 55' W) is the capital of a municipio with the same name, belonging to the Prov. Larecaja in the Dpto. La Paz, Bolivia. About 40 km east of Guanay the canton Santa Ana de Alto Beni is located, which lies about 30 km downstream the union of Río Boopi in Río Beni.

When RUSBY (1910), based on the collection *R.S. Williams 738*, described *Catopsis gracilis*, he did not chose a holotype from among the three specimens. Therefore, in accordance with the ICBN, Art. 9.9 (MCNEILL et al. 2006) the designation of the specimen preserved in [NY] by SMITH & DOWNS (1974) as holotype should be changed to lectotype.

Etymology

The epithet refers to the gracile appearance of the inflorescence with its very slender branches.

Description

Plant acaulescent, up to 40 cm high. **Leaves** many, 15–30, forming a dense, flat rosette. **Sheaths** to 30–40 mm wide, entire, whitish, the outer ones villous. **Blades** narrowly-triangular, to 40 cm long and 2.5–3.5 cm wide, thin, entire, light green, adaxially glabrous, abaxially sparsely tomentose by sessile, stellate trichomes. **Peduncle** 15–35 cm long, 2–4 mm in diameter, glabrous, glaucous. **Peduncle bracts** 0.5–2.5 cm long, shorter than or equalling the internodes, entire, abaxially villous. **Inflorescence** paniculate with lateral branches up to 2nd order, 20–35 cm long and 10–30 cm wide, axes green, glabrous, glaucous. **Primary bracts** 5–10 mm long, shorter than the sterile base of the branches, entire, abaxially sparsely villous, glabrescent. **Branches** 10–15, 8–25 cm long, inclined, arcuate, bearing up to 25 flowers. **Secondary branches** 5–10 cm long, bearing up to 15 flowers. **Floral bracts** 1–2 mm long, equalling the pedicels, entire, abaxially glabrous. **Flowers** secund, pendulous, 3–5 mm apart. **Pedicels** 1–2 mm long. **Sepals** 2–3 mm long, green, glabrous, glaucous. **Petals** 6–8 mm long, yellow, recurved during anthesis, straight afterwards. **Filaments** 4 mm long. **Anthers** 2 mm long. **Style** 3–4 mm long. **Stigmatic complex** conduplicate-spiral. **Capsule** ovoid, 3–4 mm long, 2–3 mm wide. **Seeds** filiform, bicaudate, 2 mm long.

Specimens seen

BOLIVIA: **Dpto. La Paz:** Prov. Franz Tamayo: Area Natural de Manejo Integrado Madidi; Unapa, 21 km N of Apolo, 14° 32' 26" S, 68° 29' 46" W, 1022 m, 1 Sept. 2004, *Fuentes & Aldana 6371* [LPB!]; ibid.: Parque Nacional Madidi, track between Virgen del Rosa Río and Pata, 14° 35' 48" S, 68° 40' 59" W, 1150 m, 10 Nov. 2003, *Fuentes & Cuevas 5887* [LPB!]. Prov. Sud Yungas: Near Santa Ana, next to Río Boopi, (15° 54' S, 67° 13' W), 500 m, 27 Sept. 1921, *White 1086* [K!, NY!]. **Dpto. Beni:** Prov. Ballivian: Along Río Beni, 5 km upstream from Rurrenabaque, 14° 33' 34" S, 67° 30' 35" W, 223 m, 28 Oct. 2002, *Rex & Schulte 281002-1* [SEL!]; ibid.: Along Río Beni, 5 km upstream from Rurrenabaque, 14° 33' 34" S, 67° 30' 35" W, 223 m, 28 Oct. 2002, *Rex & Schulte 281002-2* [LPB!]; ibid.: Along Río Beni, 5 km upstream from Rurrenabaque, 14° 33' 34" S, 67° 30' 35" W, 223 m, 28 Oct. 2002, *Rex & Schulte 281002-3* [LPB!, SEL!; BGHD!]; ibid.: Along Río Beni, 5 km upstream from Rurrenabaque, 14° 33' 34" S, 67° 30' 35" W, 223 m, 28 Oct. 2002, *Rex & Schulte 281002-4* [VASQ!; FAN!]; ibid.: Along Río Beni, 5 km upstream from Rurrenabaque, 14° 33' 34" S, 67° 30' 35" W, 223 m, 28 Oct. 2002, *Rex & Schulte 281002-5* [VASQ!; FAN!]; ibid.: Along Río Beni, 5 km upstream from Rurrenabaque, 14° 33' 34" S, 67° 30' 35" W, 223 m, 28 Oct. 2002, *Rex & Schulte 281002-6* [FAN!]; ibid.: Rurrenabaque, waterfall of Bala, 14° 32' S, 67° 29' W, 280 m, 28 Oct. 2002, *Vásquez 4682* [VASQ!; FAN!]; ibid.: Rurrenabaque, Río Beni, 14° 26' 32" S, 67° 31' 50", 200 m, 9 Oct. 2002, *Fuentes 5456* [LPB!, MO!].

Distribution and ecology

Range size:	Small.
Countries:	BOLIVIA. Dpto. La Paz, Beni.
Ecoregions:	Sub- and Pre-Andean-Amazon-Forest.
Life style & habitats:	Humid, evergreen lowland and montane forests of the southwest Amazon. Exposed, sunny and humid rocks on a riverbank (*Rex & Schulte 281002-3* [LPB]). Subandean forest with *Anadenanthera colubrina* (Fabaceae) (*Fuentes & Aldana 6371* [LPB]).
Altitude:	200–500 m.

Taxonomic delimitation and systematic relationships

Fosterella gracilis superficially is similar to *F. penduliflora*, but differs in the yellow petals, abaxially villous leaf-blades and very slender inflorescence with curved-ascending branches. This taxon recently has been re-established (IBISCH et al. 2006), after synonymisation by SMITH & DOWNS (1992) under *Fosterella penduliflora*. It was originally described as *Catopsis gracilis*, and temporarily shifted to *Lindmania*. The neglect of information about the colour of the petals – provided by the collector and included in the original description – by later authors led to taxonomic confusion. *Fosterella gracilis* is the only known yellow-flowering species in the genus. Evidence from molecular studies (REX et al. 2009) confirms the genetic distinctness of *F. gracilis* from *F. penduliflora*. In the phylogeny inferred from AFLP-analysis, *F. gracilis* has an isolated position, not assignable to any of the species groups. In the tree based on the DNA-sequence-data of four chloroplast loci, *F. gracilis* is sister to the widely distributed *F. penduliflora* with high levels of support.

Fig. 47: Lectotype of *Fosterella gracilis* (Rusby) L.B. Sm. [NY].

Fig. 48: *F. gracilis (Rex & Schulte 281002-3)* in the natural habitat. Photo: M. Rex.

Fig. 49: Flowers of *F. gracilis (Rex & Schulte 281002-3)*.

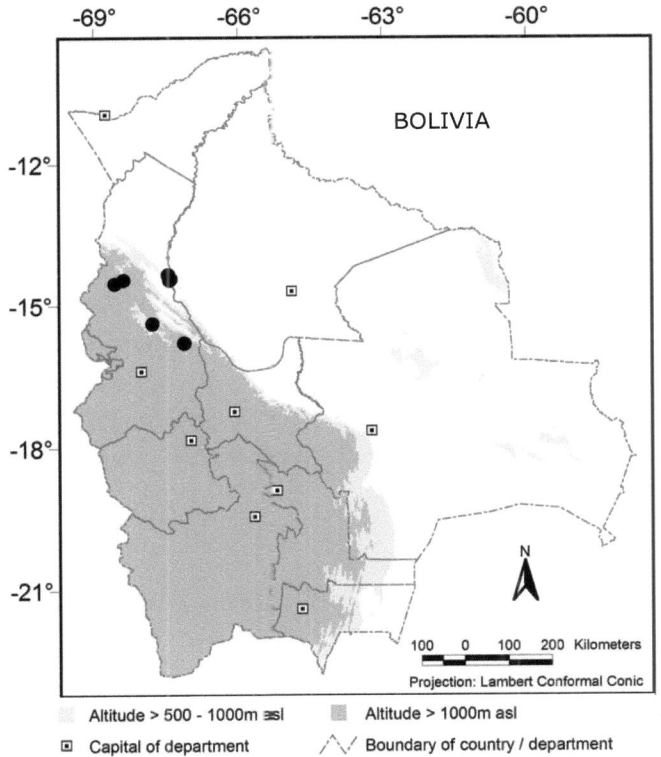

Fig. 50: Distribution of *Fosterella gracilis*.

Fosterella graminea (L.B. Sm.) L.B. Sm., Phytologia 7: 171. 1960. Fig. 51–53.

Basionym: *Lindmania graminea* L.B. Sm., Lilloa 14: 93. 1948. TYPE: Bolivia. Dpto. La Paz: Region of Mapiri, San Carlos, 700 m, 15 Nov. 1926, *O. Buchtien 417* [Holotype: US (phot!); Isotype: GH (phot!)].

Comment on type
The city Mapiri is situated in the Province Larecaja at approx. 15° 15' S, 68° 10' W.

Etymology
The epithet refers to the narrowly linear, grass-like leaf blades and peduncle bracts.

Description
Plant acaulescent, up to 100 cm high. **Leaves** many, 20–30, forming an open, more or less upright rosette. **Sheaths** 20–30 mm wide, entire, whitish, glabrous. **Blades** linear, filiform-acuminate, narrowed towards the base but not petiolate, strongly involute, to 60 cm long and 1.2 cm wide, succulent and serrate towards the base, adaxially glabrous, abaxially densely appressed lepidote by peltate trichomes with fimbriate margin. **Peduncle** 30–50 cm long, to 5 mm in diameter, green, glabrous. **Peduncle bracts** very large, subfoliaceous, 10–25 cm long, much longer than the internodes, entire, abaxially densely appressed lepidote. **Inflorescence** paniculate with branches up to 2nd order, to 60 cm long and 20 cm wide, axes green, glabrous. **Primary bracts** 30–50 mm long, longer than the sterile base of the branches, entire, abaxially appressed lepidote, glabrescent. **Branches** 20–30, inclined, arcuate, 10–15 cm long, bearing up to 50 flowers. **Secondary branches** 5–8 cm long, bearing up to 15 flowers. **Floral bracts** 1 mm long, shorter than the pedicels, entire, abaxially glabrous. **Flowers** secund, pendulous, 3–5 mm apart. **Pedicels** to 5 mm long. **Sepals** 1.5 mm long, green, glabrous. **Petals** 3 mm long, white, recoiled like watchsprings during anthesis and afterwards. **Filaments** 3 mm long. **Anthers** 2 mm long. **Style** 3 mm long. **Stigmatic complex** simple-erect. **Capsule** ovoid, 4–5 mm long, 3 mm wide. **Seeds** filiform, bicaudate, 3 mm long.

Specimens seen
BOLIVIA: **Dpto. La Paz:** Prov. Larecaja: Between Chima and Llipi, 15° 35' S, 68° 09' W, 850 m, 9 Jan. 2001, *Müller 22* [USZ!]; ibid.: Between Chima and Llipi, 15° 35' S, 68° 09' W, 850 m, 9 Jan. 2001, *Müller 7* [LPB!]; ibid.: Below Chusi, 14 Jan. 2001, 15° 37' S, 68° 15' W, 1200 m, *Müller 216* [SEL!].
WITHOUT LOCALITY: *s.n.* [WU 7466!, 9130!; HBV B312/96 ex living collection Dills, S.L. Obtopo, California].

Distribution and ecology

Range size: Very small.
Country: BOLIVIA. Dpto. La Paz.
Ecoregions: Sub-Andean-Amazon-Forests, Yungas.
Life style & habitats: Humid, evergreen lowland and montane forests.
Altitude: 700–1200 m.

Taxonomic delimitation and systematic relationships

Morphologically, *F. graminea* is similar to *F. kroemeri*, but has more, longer and narrower leaf blades, very long, graminoid peduncle bracts and larger inflorescences with longer primary bracts and pedicels. According to molecular data based on chloroplast-DNA-sequence-data (REX et al. 2009), *F. graminea* forms a lineage together with *F. albicans*, *F. caulescens*, *F. rexiae*, *F. kroemeri*, *F. robertreadii* and *F. heterophylla*, though the group is highly heterogeneous in morphological respect.

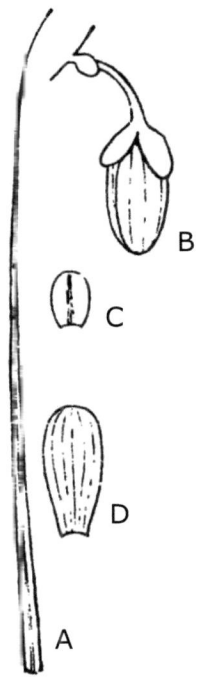

Fig. 51: *Fosterella graminea*. A Leaf. B Flower. C Sepal. D Petal. Drawing of *Buchtien 417* in SMITH (1948).
© Instituto Miguel Lillo.

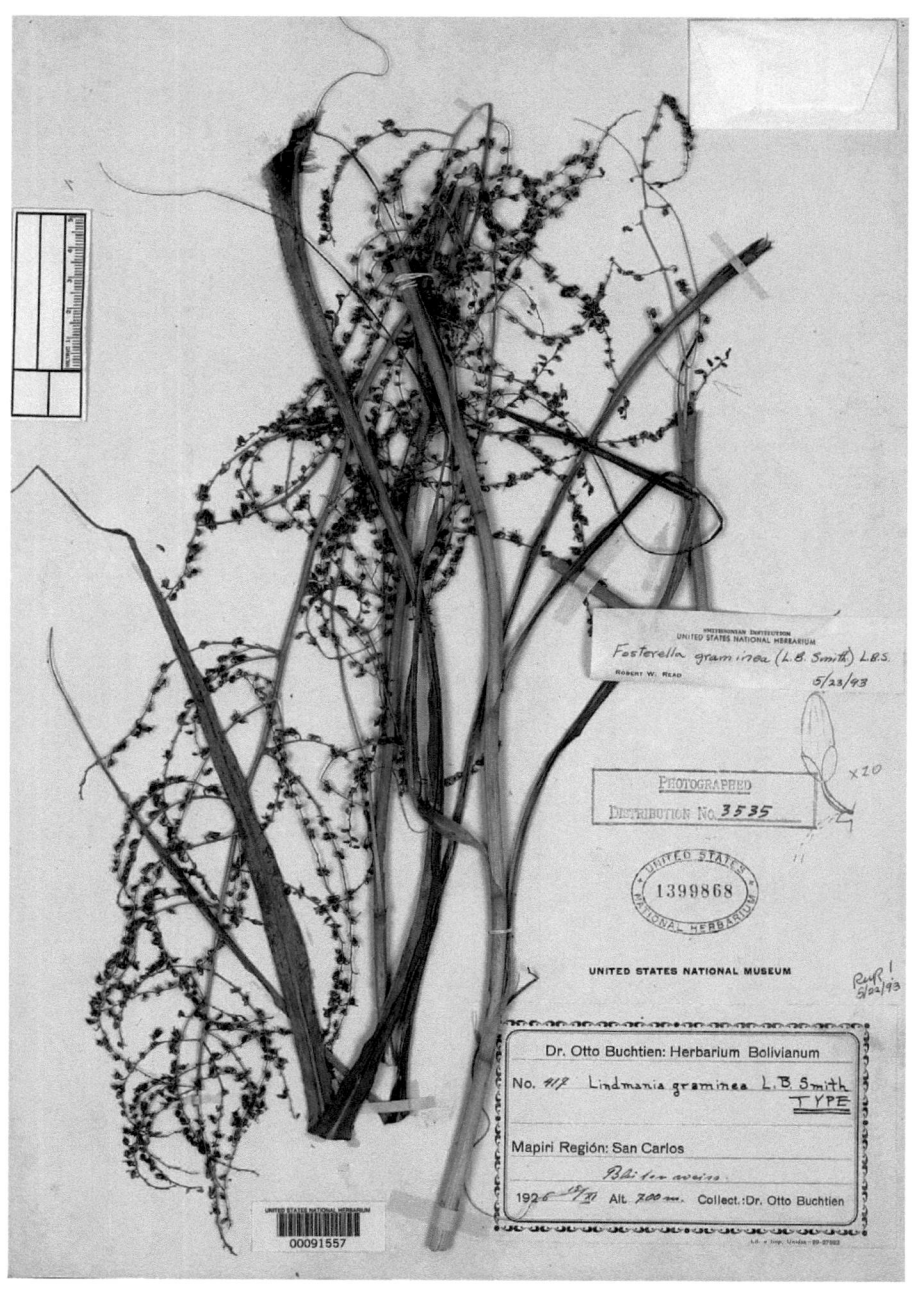

Fig. 52: Holotype of *Fosterella graminea* (L.B. Sm.) L.B. Sm. [US].
© Smithsonian National Museum of Natural History.

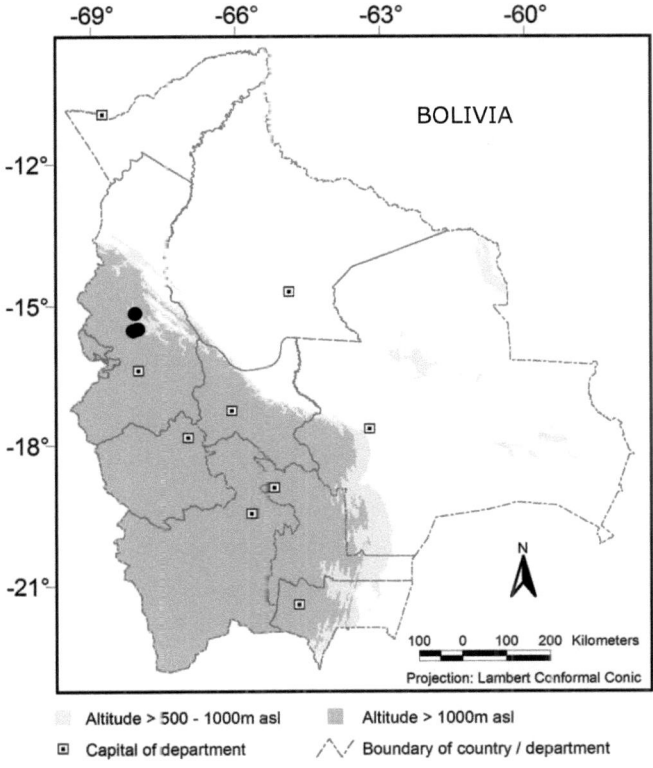

Fig. 53: Distribution of *Fosterella graminea*.

Fosterella hatschbachii L.B. Sm. & Read, Bradea 6 (15): 137. 1992. Fig. 54–57.

TYPE: Brazil. Estado Mato Grosso do Sul: Município de Aquidauana, Piraputanga, 16 Oct. 1972, *G. Hatschbach 30497* [Holotype: US (phot.!); Isotype: HB!, MBM (phot.!), NY!].

Comment on type

The type locality probably is placed approx. at 20° 30' S, 55° 48' W.

Etymology

This species is dedicated to Gert Hatschbach (*1923), a Brazilian botanist and collector of the type specimen.

Description

Plant acaulescent, up to 40 cm high. **Leaves** few, 12–15, forming an open, flat rosette. **Sheaths** to 25 mm wide, entire, whitish, glabrous. **Blades** linear, acuminate, narrowed towards the base, 10–50 cm long and 1–2.5 cm wide, thin, entire, glabrous, abaxially densely tomentose by sessile, stellate trichomes. **Peduncle** (5) 10–20 cm long, 1–2 mm in diameter, green, glabrous. **Peduncle bracts** 1–2 cm long, shorter than or equalling the internodes, entire, abaxially villous. **Inflorescence** racemose or compound racemose, rarely paniculate with branches up to 2nd order, (5) 10–20 cm long and 5–15 cm wide, axes green, glabrous. **Primary bracts** 3–7 mm long, shorter than the sterile base of the branches, entire, abaxially sparsely villous. **Branches** up to 10, inclined, arcuate, to 8 cm long, bearing up to 12 flowers. **Secondary branches** up to 3 cm long, bearing up to 5 flowers. **Floral bracts** 2 mm long, shorter than or equalling the pedicels, entire, abaxially glabrous. **Flowers** secund, pendulous, 3 mm apart. **Pedicels** 2–3 mm long. **Sepals** 2–3 mm long, green, glabrous. **Petals** 7 mm long, white, recurved during anthesis, straight afterwards. **Filaments** 4 mm long. **Anthers** 2 mm long. **Style** 5 mm long. **Stigmatic complex** simple-erect. **Capsule** narrowly ovoid, 5 mm long, 2 mm wide. **Seeds** filiform, bicaudate, 2 mm long.

Specimens seen

BRAZIL: **Estado Mato Grosso:** Mun. Chapada dos Guimarães: Without precise locality, 7 Nov. 1983, *Kautsky 830* [HB!]; ibid.: 1980, *Lima 13* [HB!]; ibid.: 12 Nov. 2007, *Leme 7096* [LEME!]; ibid.: 12 Nov. 2007, *Leme 7100* [LEME!, BGHD!]; ibid.: Along Sete de Setembro River, on vertical walls at the riverbank, 780 m, 6 Nov. 2006, *Leme & Gonzalez 6970* [LEME!]; ibid.: 5 km E from the town of Chapada dos Guimarães towards Embratal, 720 m, 24 Oct. 1973, *Prance 19368* [K!, MBM (phot!), NY!]; ibid.: Next to Agua Fria, 20 Oct. 1995, *Hatschbach 63638* [HBG!, M!, MA!, NY!]; ibid.: Next to Véu de Noiva, along Río Coxipozinho, 15° 30' S, 55° 45' W, 21 Oct. 1985, *Pirani 1313* [NY!]; ibid.: Portão do Inferno, 16 Nov. 1975, *Hatschbach 37652* [MBM (phot!)]. **Estado Mato Grosso do Sul:** Mun. Aquidauana: Piraputanga, 6 Dec 1970, *Hatschbach 25757* [MBM (phot!)].

Distribution and ecology

Range size: Very small.
Countries: BRAZIL. Estado Mato Grosso, Mato Grosso do Sul.
Ecoregion: Cerrado Forests.
Life style & habitats: On Precambrian rock outcrops. Terrestrial or saxicolous, growing on riverbanks and in shady understory of more or less disturbed, deciduous forests. Between bryophytes and small pteridophytes, quite rare (*Pirani 1313* [NY]).
Altitude: 700 m.

Taxonomic delimitation and systematic relationships

Fosterella hatschbachii is similar to *F. penduliflora* but differs in the densely tomentose abaxial leaf surfaces and in ecological requirements: *Fosterella hatschbachii* is restricted to Precambrian rocks within the Cerrado Forests of Mato Grosso, Brazil, whereas the widely distributed *F. penduliflora* occurs in the montane forests of Bolivia.

Furthermore *F. hatschbachii* resembles *F. yuvinkae*, but differs in smaller stature; less number of leaves, shorter peduncle, less branched inflorescence, shorter primary bracts, branches, floral bracts and petals. Moreover *F. yuvinkae* differs from all other *Fosterella* species by its very narrow petals.

Hitherto no molecular data is available for *F. hatschbachii*

Some specimens from Chapada dos Guimarães are remarkably small (*Hatschbach 37652, Prance 19368, Leme & Gonzales 6970, Leme 7096, Leme 7105*), which has led to the assumption that those specimens might represent a new species (LEME, pers.comm.). However it is absolutely neccessary to study the holotype locality to be able to come to a decision.

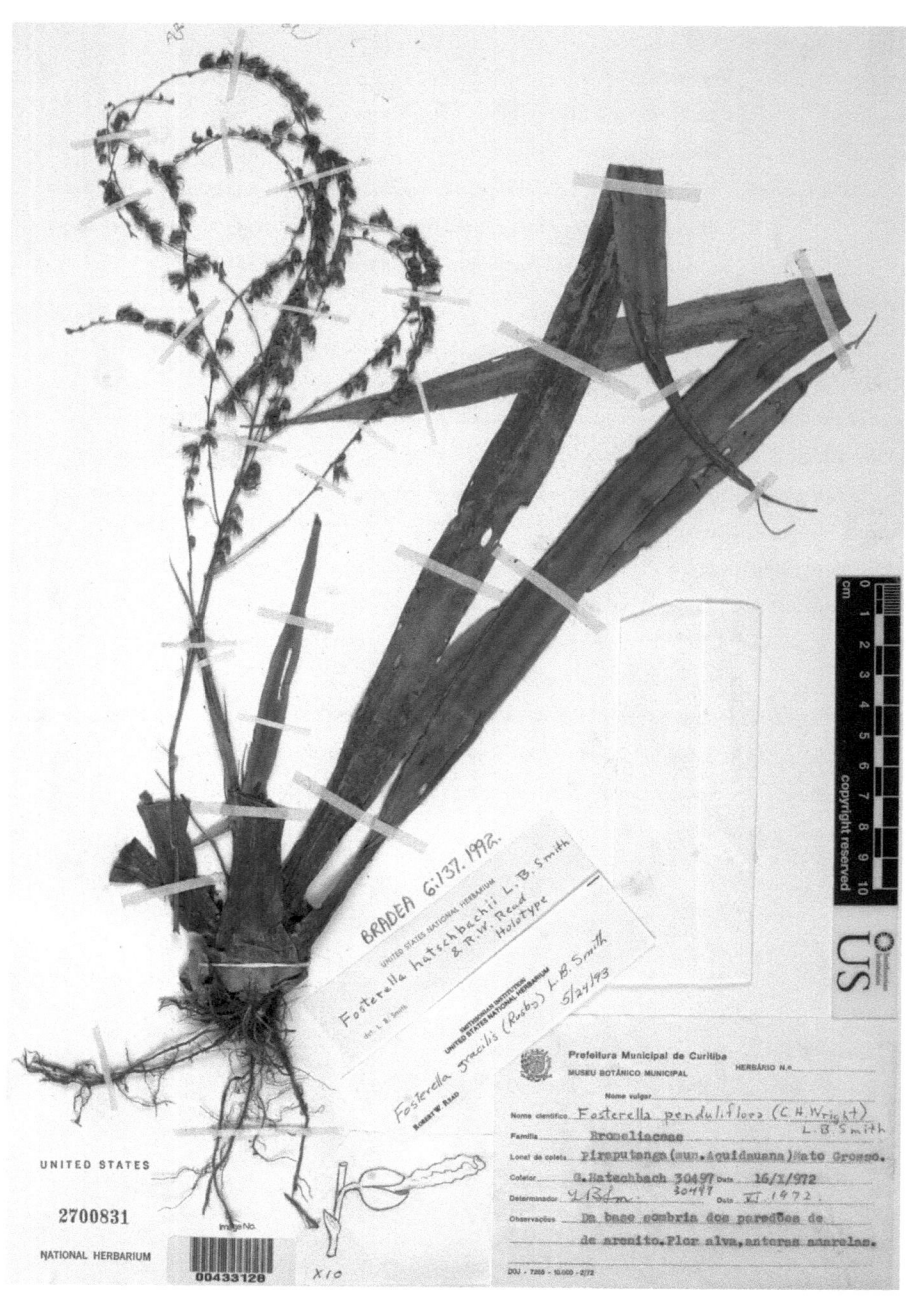

Fig. 54: Holotype of *Fosterella hatschbachii* L.B. Sm. & Read [US].
© Smithsonian National Museum of Natural History.

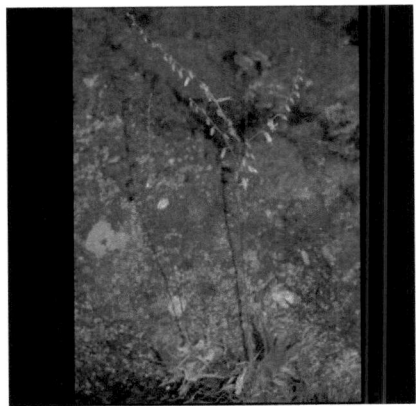

Fig. 55: *F. hatschbachii (Leme 7095)* in the natural habitat. Photo: E. Leme.

Fig. 56: Flowers of *F. hatschbachii (Leme 7096)*. Photo: E. Leme.

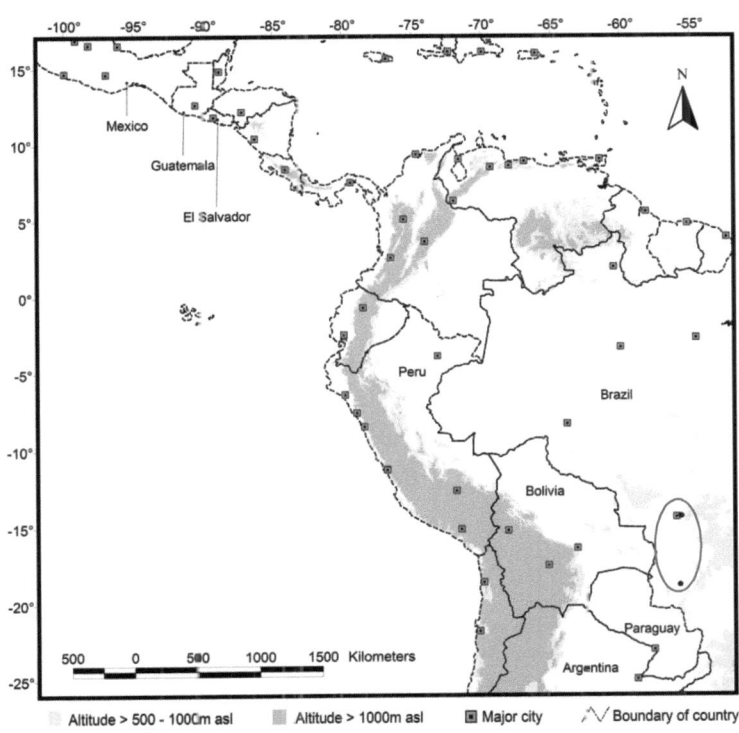

Fig. 57: Distribution of *Fosterella hatschbachii*.

Fosterella heterophylla Rauh, Trop. Subtrop. Pflanzenwelt 60: 24. 1987. Fig. 58–62.

TYPUS: Bolivia. Yungas, close to La Paz: 1000 m, 16 Aug. 1976, *W. Rauh 40583 a* [Holotype: HEID!].

Comment on type

The type locality is situated in the La Paz Department of Bolivia, unfortunately not well specified.

Etymology

The epithet refers to the dimorphic foliage of this species.

Description

Plant caulescent, up to 60 cm high. **Leaves** dimorph; lower ones narrowly triangular, surrounding the stem, gradually merging into the rosette leaves, which are few, up to 10, distinct petiolate, forming an open, flat rosette. **Sheaths** to 30 mm wide, entire, reddish, glabrous. **Blades** petiolate, lanceolate, acuminate, to 20 cm long and 3 cm wide, thin, entire, (petioles 8–10 cm long, succulent towards the base, entire), adaxially glabrous, abaxially densely appressed lepidote by peltate trichomes with fimbriate margin. **Peduncle** to 30 cm long, 5 mm in diameter, frequently reddish, glabrous. **Peduncle bracts** 1.5–2.5 cm long, longer than or equalling the internodes, entire, frequently reddish, abaxially densely appressed lepidote. **Inflorescence** compound racemose, rarely paniculate with lateral branches of 2nd order, to 35 cm long and 17 cm wide, axes frequently reddish, glabrous. **Primary bracts** to 10 mm long, longer than the sterile base of the branches, entire, frequently reddish, abaxially appressed lepidote. **Branches** 5–15, ascending, straight, 5–10 cm long, bearing up to 20 flowers. **Secondary branches** 4–5 cm long, bearing 5–7 flowers. **Floral bracts** 3–6 mm long, longer than or equalling the pedicels, entire, abaxially sparsely appressed lepidote. **Flowers** secund, pendulous, 3–5 mm apart. **Pedicels** 2–4 mm long. **Sepals** 2–3 mm long, glabrous. **Petals** 5 mm long, white, recoiled like watchsprings during anthesis and afterwards. **Filaments** 4 mm long. **Anthers** 2 mm long. **Style** 3 mm long. **Stigmatic complex** conduplicate-spiral. **Capsule** narrowly ovoid, 5 mm long, 2 mm wide. **Seeds** filiform, bicaudate, 2 mm long.

Specimens seen

BOLIVIA: **Dpto. La Paz:** Prov. Caranavi: 34 km from Caranavi to Yucumo, 15° 41' S, 67° 29' W, 1600 m, 14 Feb. 2000, *Vásquez & Gerlach 3661* [FR!, LPB!, SEL!; FAN!]. Nor Yungas: Between Viscachani and Coroico, 16° 14' 29" S, 67° 47' 24" W, 2152 m, 24 Oct. 2002, *Rex & Schulte 241002-1* [FR!]; ibid.: Between Viscachani and Coroico, 16° 14' 29" S, 67° 47' 24" W, 2152 m, 24 Oct. 2002, *Rex & Schulte 241002-2* [SEL!]; ibid.: Between Viscachani and Coroico, 16° 14' 29" S, 67° 47' 24" W, 2152 m, 24 Oct. 2002, *Rex & Schulte 241002-3* [LPB!]; ibid.: Between Viscachani and Coroico, 16°14' 29" S, 67° 47' 24" W, 2152 m, 24 Oct. 2002, *Rex & Schulte 241002-5* [FR!; FAN!]; ibid.: Between Viscachani and Coroico, 16° 14' 29" S, 67° 47' 24" W, 2152 m, 24 Oct. 2002, *Rex & Schulte 241002-6* [USZ!];

Between Viscachani and Coroico, 16° 14' 29" S, 67° 47' 24" W, 2152 m, 24 Oct. 2002, *Rex & Schulte 241002-7* [SEL!].
WITHOUT LOCALITY: *Rauh s.n.* [B!; BGB 289-51-00-63! ex HBV].

Distribution and ecology

Range size:	Very small.
Country:	BOLIVIA. Dpto. La Paz.
Ecoregion:	Yungas.
Life style & habitats:	Humid, evergreen montane forests. Terrestrial or saxicolous. On steep slopes, between bryophytes in shady understory (*Rex & Schulte 241002-2* [SEL]).
Altitude:	100–2150 m.

Taxonomic delimitation and systematic relationships

Fosterella heterophylla is similar to *F. weddelliana*, but differs in petiolate and shorter leaf blades, shorter peduncle, primary bracts and floral bracts.

According to molecular data based on chloroplast-DNA-sequence-data (REX et al. 2009), *F. heterophylla* forms a clade together with *F. albicans*, *F. caulescens*, *F. rexiae*, *F. kroemeri*, *F. graminea* and *F. robertreadii*, which morphologically is a highly heterogeneous group.

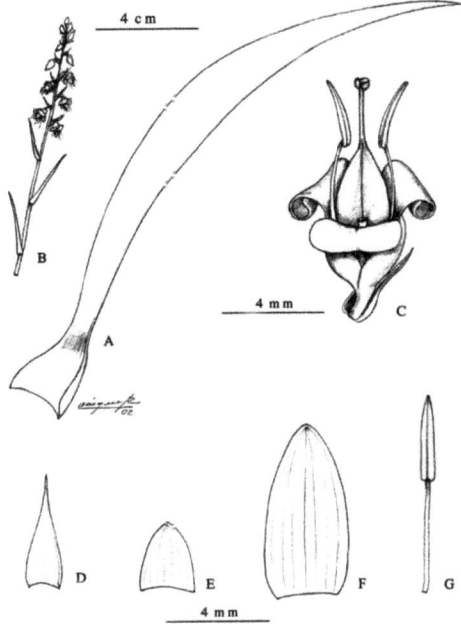

Fig. 58: *Fosterella heterophylla*. A Leaf. B Inflorescence detail. C Flower. D Floral bract. E Sepal. F Petal. G Stamen. Drawing of *Vásquez & Gerlach 3661* by R. Vásquez in IBISCH et al. (2002). © Selbyana.

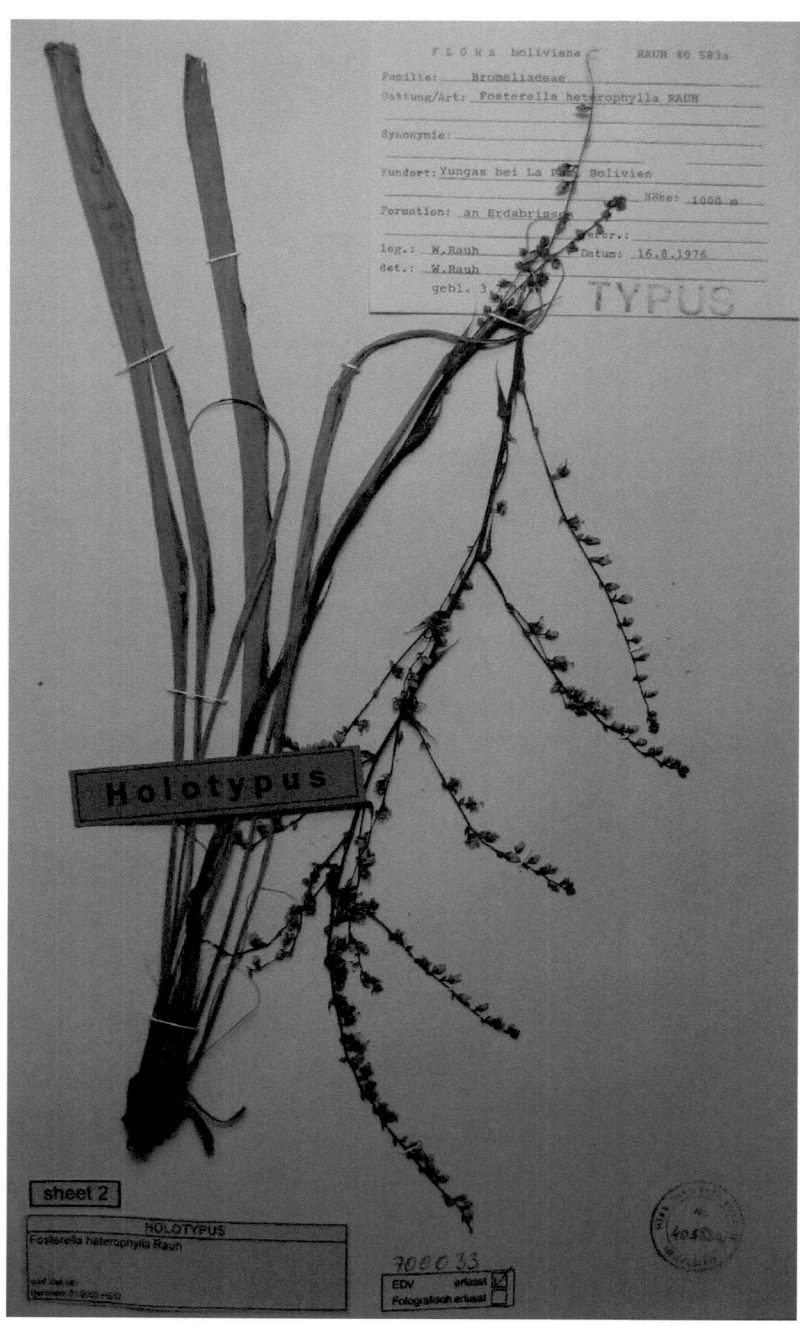

Fig. 59: Holotype of *Fosterella heterophylla* Rauh [HEID].

Fig. 60: *F. heterophylla (Rauh s.n.)*.

Fig. 61: Flowers of *F. heterophylla (Vásquez & Gerlach 3661)*. Photo: P. Ibisch.

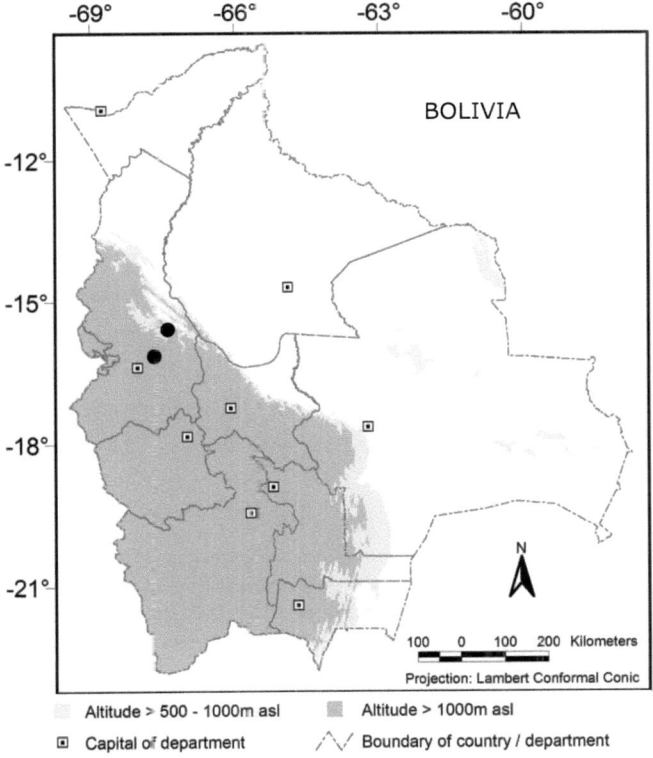

Fig. 62: Distribution of *Fosterella heterophylla*.

Fosterella kroemeri Ibisch, R. Vásquez & J. Peters, Selbyana 29 (2): 189. 2008. Fig. 63–67.

TYPE: Bolivia. Dpto. La Paz: Prov. Caranavi, road from Caranavi to Sapecho, near the summit of the Serranía Bella Vista, 15° 41' S, 67° 29' W, 1500 m, *T. Krömer & A. Aceby 1398 b* [Holotype: LPB!; Isotype: SEL!].

Etymology

The species is dedicated to the collector of the type, Thorsten Krömer (*1969), German bromeliad specialist and epiphyte ecologist who has significantly contributed to the knowledge of the bromeliads of Bolivia.

Description

Plant acaulescent, up to 60 cm high. **Leaves** few, 10 to 15, forming an open, more or less upright rosette. **Sheaths** to 30 mm wide, entire, whitish, glabrous. **Blades** oblanceolate, narrowed towards the base, to 50 cm long and 3 cm wide, succulent towards the base, entire or with minute spines at the base, adaxially glabrous, abaxially densely appressed lepidote by peltate trichomes with fimbriate margin. **Peduncle** to 40 cm long, 4–5 mm in diameter, green, glabrous. **Peduncle bracts** to 3 cm long, equalling the internodes, entire, sometimes reddish, abaxially appressed lepidote. **Inflorescence** compound racemose, rarely paniculate with lateral branches of 2nd order, to 25 cm long and 8 cm wide, axes green, glabrous, glaucous. **Primary bracts** to 8 mm long, equalling the sterile base of the branches, entire, sometimes reddish, abaxially appressed lepidote. **Branches** up to 20, ascending, straight, to 5 cm long, zigzag-shaped, bearing up to 15 flowers. **Secondary branches** 2–3 cm long, bearing 3–5 flowers. **Floral bracts** 1–2 mm long, shorter than the pedicels, entire, sometimes reddish, abaxially lepidote, glabrescent. **Flowers** subsecund, pendulous, 15 mm apart. **Pedicels** 2–3 mm long. **Sepals** 1 mm long, green, glabrous. **Petals** 4 mm long, white, recoiled like watchsprings during anthesis and afterwards. **Filaments** 4 mm long. **Anthers** 1.5 mm long. **Style** 3 mm long. **Stigmatic complex** simple-erect. **Capsule** ovoid, 5 mm long, 3 mm wide. **Seeds** filiform, bicaudate, 2 mm long.

Specimens seen

BOLIVIA: **Dpto. La Paz:** Prov. Nor Yungas: Between Viscachani und Coroico, 16° 13' 31" S, 67° 45' 12" W, 1420 m, 24 Oct. 2002, *Rex & Schulte 241002-12* [FR!, SEL!; FAN!]; ibid.: Between San Pedro and Chojna, ca. 1–2km before Chojna, 15° 59' 14" S, 67° 35' 23" W, 923 m, 26 Oct. 2002, *Rex & Schulte 261002-10* [FR!]; ibid.: Between San Pedro and Chojna, ca. 1–2 km before Chojna, 15° 59' 14" S, 67° 35' 23" W, 923 m, 26 Oct. 2002, *Rex & Schulte 261002-11* [SEL!]; ibid.: 25 km from Caranavi to Yolosa, 15° 57' S, 67° 34' W, 830 m, 15 Feb. 2000, *Vásquez & Gerlach 3674 a* [SEL!]; ibid.: Between Chuspipata and Yolosa, 16° 13' S, 67° 44' W, 1970 m, 13 Feb. 2000, *Vásquez & Gerlach 3652* [LPB!, SEL!]; ibid.: 46 km from Yolosa to Caranavi, 15° 59' S, 67° 35' W, 800 m, 26 Oct. 2002, *Vásquez 4654 a* [LPB!].

Distribution and ecology

Range size: Very small.

Country: BOLIVIA. Dpto. La Paz.

Ecoregion: Yungas.

Life style & habitats: Humid, evergreen montane forests. Terrestrial or saxicolous on steep, rocky slopes, between bryophytes in shady understory (*Rex & Schulte 261002-11* [SEL]).

Altitude: 1400–1500 m.

Taxonomic delimitation and systematic relationships

Fosterella kroemeri is quite similar to *F. windischii* but differs in longer and wider leaf blades whith less pronounced indumentum on the abaxial leaf surface, longer peduncles and more, but shorter inflorescence-branches.

Molecular results from comparative sequence-analysis of four chloroplast DNA loci (REX et al 2009) show that the two specimens are not closely related with each other. According to the molecular trees, *F. windischii* belongs to a group growing on Precambrian outcrops that also includes *F. vasquezii* and *F. yuvinkae*, while *F. kroemeri* groups together with *F. caulescens, F. rexiae* and *F. albicans*. The ecological differences with *F. kroemeri* occuring in the humid montane rain forests of Bolivia and *F. windischii* being confined to a small area of Precambrian rocks of the Brazilian shield give further support for taxonomical separation.

Two specimens which formerly were assigned to *F. rojasii*, (*Vásquez & Gerlach 3674 a* [SEL!] and *Vásquez & Gerlach 3652* [LPB!, SEL!]) now have to be identified as *F. kroemeri*, despite their somewhat aberrant habitus showing a rather flat rosette.

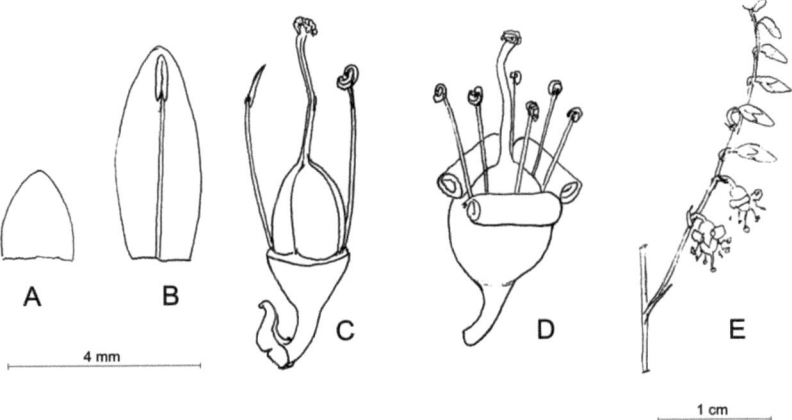

Fig. 63: *Fosterella kroemeri*: A Sepal. B Petal and stamen. C Gynoecium and stamen. D Flower. E Inflorescence branch. Drawing of *Krömer & Acebey 1398 b* by R. Vásquez in PETERS et al. (2008 b). © Selbyana.

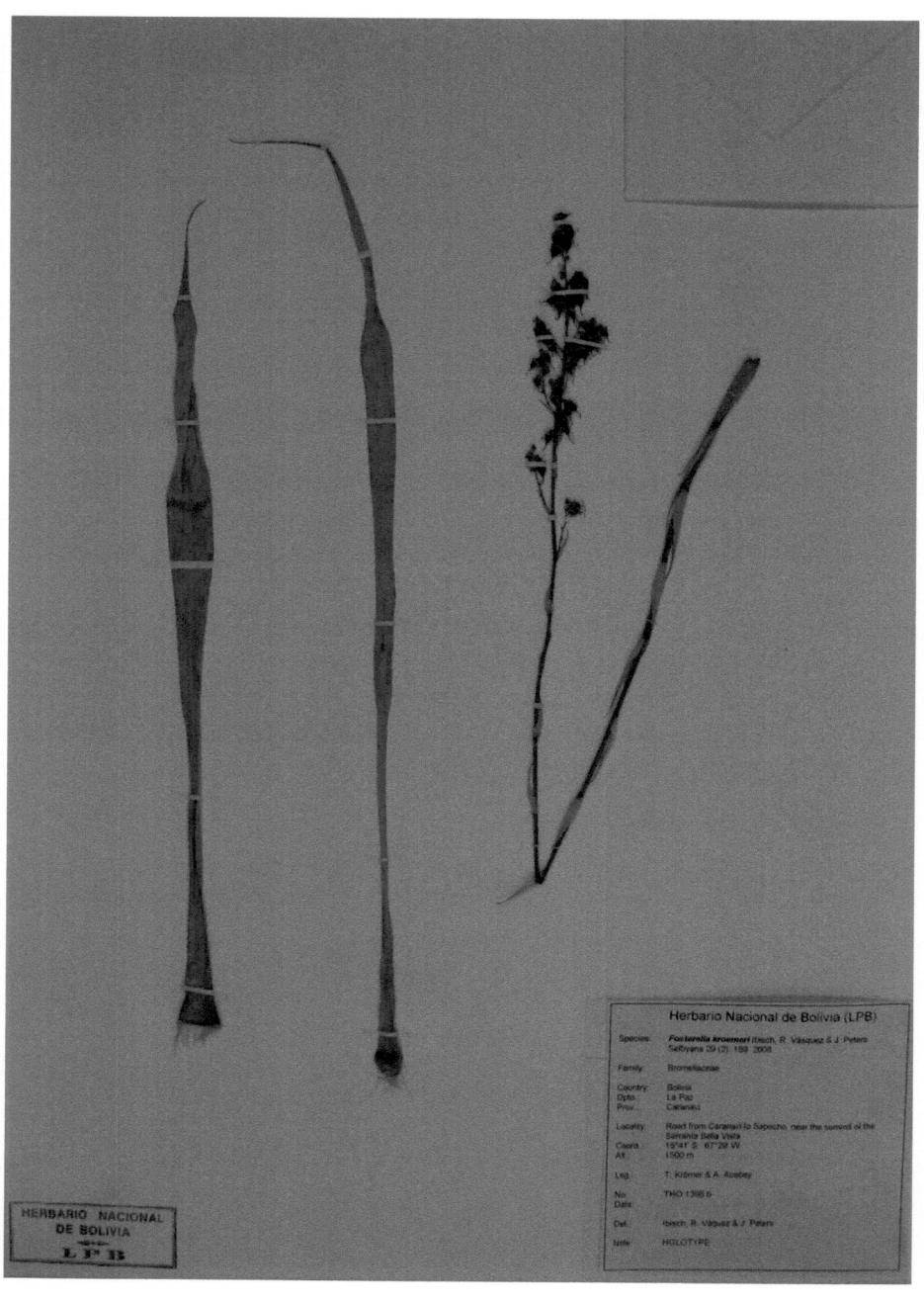

Fig. 64: Holotype of *Fosterella kroemeri* Ibisch, R. Vásquez & J. Peters [LPB].

Fig. 65: *F. kroemeri (Krömer & Acebey 1398 b)*. Photo: P. Ibisch.

Fig. 66: Flowers of *F. kroemeri (Krömer & Acebey 1398 b)*.

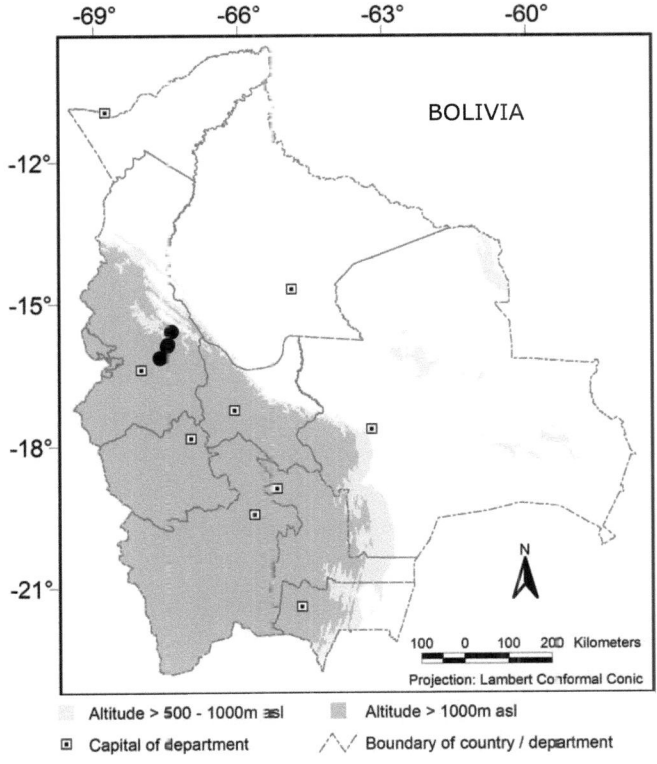

Fig. 67: Distribution of *Fosterella kroemeri*

Fosterella micrantha (Lindl.) L.B. Sm., Phytologia 7: 171. 1960. Fig. 68–72.

Basionym:	*Pitcairnia micrantha* Lindl., Edwards's Bot. Reg. 29: Misc. 44. 1843. TYPE: Found among some Orchidaceous plants imported from Rio, in December, 1841, by Lieut. Christopher Smith, of H.M. Packet "Star", and presented to Sir Charles Lemon, Bart. M.P. [Holotype: CGE (phot.!)].
≡	*Lindmania micrantha* (Lindl.) L.B. Sm., Contr. Gray Herb. 104: 77. 1934.
=	*Cottendorfia neogranatensis* Baker, Handbook of the Bromeliaceae: 129. 1889. TYPE: New Granada, *C. Jürgensen 389* [Holotype: K (phot.!)].
≡	*Lindmania neogranatensis* (Baker) Mez in C. DC., Monogr. Phan. 9: 538. 1896.
=	*Lindmania flaccida* Standl., J. Wash. Acad. Sci. 13: 364. 1923. TYPE: El Salvador. Dpto. Ahuachapán: Moist shaded bank along a stream in the mountains near Ahuachapán, 800–1000 m, Jan. 1922, *P.C. Standley 19786* [Holotype: US (phot.!); Isotypes: NY!, GH (phot.!)].

Comment on type

LINDLEY (1843) described *Pitcairnia micrantha* Lindl. as being imported from Rio (which most likely means Rio de Janeiro, Brazil) – but this does not necessarily imply, that the plant was collected in Brazil. SMITH (1934 b) had already noticed, that a Brazilian origin of the type was very unlikely in view of subsequent collections.

Concerning the type locality of *Cottendorfia neogranatensis* Baker, the original description only says „New Granada", without further locality (BAKER 1889). The term New Granada may refer to various former national denominations formed by the present day nations of Colombia and Panama, as well as smaller portions of Venezuela, Ecuador, Perú, Brazil, Costa Rica, Nicaragua, Guayana, Trinidad and Tobago – therefore it is difficult to trace, where the plant actually comes from. Furthermore, New Granada may also mean Nueva Granada, which is a municipality in the Usultán Department of El Salvador. MEZ (1934) stated without further explanation that the type comes from the same region like Reiche's collection: Mexico, Departamento Oaxaca, Sierra Madre, Río Tanetze, *Reiche 566a*, Mar./Apr. 1927 [M]. In view of the fact, that C. Jürgensen is known for his great collection efforts in Mexico (CHAUDHRI et al. 1972), it seems most likely, that the type comes from there.

The city of Ahuachapán, next to the type locality of *Lindmania flaccida* Standl., is situated approx. at 13° 55' N, 89° 51' W.

Fosterella micrantha (Lindl.) L.B. Sm. is the type species of the genus *Fosterella* L.B. Sm.

Etymology

The epithet refers to the species' small flowers: mikros (μικρός) = small; anthos (άνθος) = flower.

Description

Plant acaulescent, up to 70 cm high. **Leaves** few, 6–12, forming an open, arched rosette. **Sheaths** 50–60 mm wide, entire, whitish, glabrous. **Blades** lanceolate, acuminate, to 35 cm long and 2–6.5 cm wide, thin, entire, light green, adaxially glabrous, abaxially scattered lepidote by peltate trichomes with dentate margin. **Peduncle** 15–40 cm long, 2–3 mm in diameter, green to slightly reddish, sparsely villous-arachnoid. **Peduncle bracts** (1) 2–4 cm long, equalling the internodes, entire abaxially sparsely villous-arachnoid. **Inflorescence** racemose or compound racemose, rarely paniculate with branches up to 2nd order, 15–35 cm long and 5–15 cm wide, axes green, sparsely villous. **Primary bracts** 10 mm long, equalling the sterile base of the branches, entire, abaxially sparsely villous. **Branches** 6–15, inclined, arcuate, 5–15 cm long, bearing up to 25 flowers. **Secondary branches** 2–4 cm long, bearing 5–8 flowers. **Floral bracts** 4–6 mm long, longer than or equalling the pedicels, entire, abaxially sparsely villous. **Flowers** secund, pendulous, 4–6 mm apart. **Pedicels** to 3 mm long. **Sepals** 3–4 mm long, green, glabrous. **Petals** 6–9 mm long, white, recurved during anthesis, straight afterwards. **Filaments** 3 mm long. **Anthers** 2 mm long. **Style** 3 mm long. **Stigmatic complex** conduplicate-spiral. **Capsule** ovoid, 4–5 mm long, 3 mm wide. **Seeds** filiform, bicaudate, 3–4 mm long.

Specimens seen

MEXICO: Without precise locality, *Utley & Utley 8150* [SEL!; MSBG]. **Dpto. Oaxaca:** Sierra Madre, (16° 30' N, 96° 00' W), Mar./Apr. 1927, *Reiche 566a* [M!]. DISTRICT CHOAPAM: Mun. Santiago Choapam: 5 km SW of Santiago Choapam, road to Totontepec, (16° 06' N, 97° 30' W), 800 m, 24 Feb. 1987, *Torres & Cortes 9302* [F!, MEXU, US!]. Mun. Santiago Yaveo: Arroyo de Culebras, (17° 18' N, 95° 42' W), 450 m, 21 Mar. 1938, *Mexia 9187a* [B!, ENCB, MEXU, MO, US, XAL]. DISTRICT IXTLÁN: Mun. Santiago Comaltepec: Between Puerto Eligio and Comaltepec, 151 km on the road from Oaxaca to Tuxtepec, (16° 19' 07" N, 97° 47' 51" W), 1000 m, 17 Mar. 1966, *Calderon 735* [MO!, NY!]. DISTRICT JUCHITÁN: Mun. Santa María Chimalapa: Without precise locality, *Hernández 798* [UAMIZ]; ibid.: Without precise locality, *Hernández 932* [UAMIZ]. DISTRICT JUQUILA: Mun. San Gabriel Mixtepec: 5 km N of San Gabriel Mixtepec, towards Puerto Escondido and Sola de Vega, (16° 05' N, 97° 06' W), 830 m, 17 Mar. 1985, *Torres & Martinez 6611* [MEXU, MO!]. DISTRICT TEHUANTEPEC: Mun. Guevea de Humboldt: N of Guevea de Humboldt, between the summit and Cerro Picacho, 16° 52' N, 95° 25' W, 1000 m, 29 Mar. 1991, *Torres 13384* [FCME, MEXU, MO!]; ibid.: Guevea de Humboldt, (16° 47' N, 95° 22' W, 620 m), *Torres, Cortés & Ramírez 9156* [IEB, MEXU]. DISTRICT POCHUTLA: Mun. San Pedro Pochutla: Without precise locality, *Rivera, Salas & Elorsa 1105* [SERO]; ibid.: Puerto Angel, (15° 44' N, 96° 29' W), 08 Feb. 1966, *Ernst 2319* [BM!, MEXU, MICH, US!]; ibid.: Vicinity of Cafetal Concordia, (15° 40' N, 96° 29' W), 500 m, Apr. 1933, *Morton & Makrinius 2395* [US!]. Mun. Candelaria Loxicha: Without precise locality, *López-Ferrari 570* [ENCB; IEB; MEXU; UAMIZ]; ibid.: Without precise locality, 2 Feb. 1945, *Alexander 453* [MICH, NY!]. Mun. San Miguel del Puerto: Without precise locality, *Cerón 355* [UAMIZ]; ibid.: San Miguel del Puerto, (15° 55' N, 96° 10' W), *Salas, Rivera & Elorsa 2696* [MEXU]; ibid.: San Miguel del Puerto, (15° 55' N, 96° 10' W), *Rivera, Salas & Elorsa 2177* [MEXU]. DISTRICT TUXTEPEC: Between Valle Nacional and Oaxaca, Vera Cruz nearby Tuxtepec, next to a waterfall, (17° 35' N, 96° 29' W), 620 m, 6 Mar. 1983, *Lautner LW 83/17* [GOET!; BGGÖ!]. Mun. San Juan Bautista Valle Nacional: Along Hwy. 175 between Valle Nacional and Oaxaca, next to bridge at Valle Nacional, 17° 44' N, 96° 19' W, 705 m, 21 Feb. 1987, *Croat & Hannon 65511* [MEXU, MO!]. Mun. San Felipe Usila: Without precise locality, *Calzada, Vargas & Ibarra 16824* [MEXU]; ibid.: Cerro Nariz, 19 Apr. 1939, *Schultes & Reko 660* [GH]. Mun. San Felipe Jalapa de Díaz: Next to Jalapa de Diaz on the road Mex 182, (18° 04' N, 96° 32' W), 500 m, 18 Feb. 2003, *Lautner L 03/40* [GOET!; BGGÖ!]. **Dpto. Guerrero:** Mun. Juan R. Escudero: La Venta, union of Río Omitlán and Río Papagayo, near Tierra Colorado, about 50 km S of Chilpancingo on the road Mex 95, (17° 08' N, 99° 32' W), 195 m, 10 Jan. 1966,

Kruse 965 [MEXU (phot.!)]; ibid.: La Venta, union of Río Omitlán and Río Papagayo, near Tierra Colorado, about 50 km S of Chilpancingo on the road Mex 95, (17° 08' N, 99° 32' W), 195 m, 13 Feb. 1966, *Kruse 1001* [MEXU (phot.!)]. **Dpto. Veracruz:** <u>Mun. Hidalgotitlán:</u> Next to Hidalgotitlán, 17° 17' 00" N, 94° 30' 45" W, 150 m, 11 Mar. 1982, *Wendt 3644* [ENCB, IEB, MEXU, MO!, NY!, UAMIZ, XAL]; ibid.: Campamento Hermanos Cedillo, banks of Río Solosúchil, *Vásquez 223* [ENCB, MEXU, XAL]; ibid.: Río Solosúchil, entre Hermanos Cedillo y La Escuadra, *V-1365* [MEXU]. <u>Mun. San Andrés Tuxla:</u> Laguna Encantada, 7 km NE of San Andres Tuxla, (18° 27' N, 95° 13' W), 26 Jan. 1973, *Cedillo & Calzada 87* [F!, NY!, MEXU]; ibid.: Laguna Encantada, 5 km NE of San Andres Tuxla, (18° 27' N, 95° 11' W), 450 m, 10 Feb. 1972, *Beaman 5639* [F!, MEXU]; ibid.: 2 km N of San Andrés Tuxla, (18° 28' N, 95° 13' W), *Cedillo 3785* [IBUG, IEB, MEXU]. <u>Mun. Soteapan:</u> Zapoapan de Cabañas, region Los Tuxlas, (18° 23' N, 95° 00' W), *Bravo 121* [MEXU]; ibid.: San Fernando, W of the village, (18° 09' N, 94° 33' W), *Leonti 432* [MEXU]. <u>Mun. Catemaco:</u> Lake Catemaco, Agaltepec Island, 18° 24' N, 95° 05' W, 5 Aug. 1976, *Faden 76/153* [F!]; ibid.: Laguna Verde, N of Catemaco, next to the road Catemaco – Playa Azul, (18° 32' N, 95° 08' W), *Calzada 11717* [XAL]. Chinameca: Pajapan, (18° 15' N, 94° 41' W) *Téllez 4469* [MEXU]. **Dpto. Chiapas:** <u>Mun. Mapastepec:</u> Sierra de Soconusco, new road from Hwy. 200 to Tuxtla Gutierrez, 15° 31' N, 92° 50' W, 200 m, 21 Jan. 1987, *Croat & Hannon 63375* [MO!]. <u>Mun. Escuintla:</u> Near Río Cintalapa, (15° 18' N, 92° 36' W), 27 Nov. 1947, *Matuda 17263* [F!, GH!, MO!, NY!]; ibid.: Next to Cintalapa, (16° 41' N, 93° 42' W), 15 Jan. 1948, *Matuda 17427* [F!, K!]; ibid.: Mount Ovando, (15° 25' N, 92° 36' W), Dec. 1937, *Matuda 2076* [GH!, K!, NY!, US!]; ibid.: Mount Ovando, (15° 25' N, 92° 36' W), 24 Oct. 1941, *Matuda 6204* [NY!]. <u>Mun. Tapachula:</u> Between San Carlos and Chiapaso, (14° 54' N, 92°17' W), 1200 m, Aug. 1974, *Rauh 36412* [HEID!]. <u>Mun. Tuzatán:</u> 15 km NE of Huixtla, (15° 16' N, 92° 23' W), 800 m, 9 Feb. 1987, *Martínez, Márquez & Urquijo 19942* [MEXU (phot.!), MO].
GUATEMALA: Without precise locality, Mar. 1961, *Palín 1797* [US!]. **Dpto. Huehuetenango:** Canyon tributary to Río Trapichillo, between Democracia and canyon of Camushú, (15° 17' N, 91° 11' W), 1050 m, 24 Aug. 1942, *Steyermark 51239* [F!]. **Dpto. San Marcos:** Finca El Porvenir, (14° 58' N, 91° 56' W), 1300 m, 24 Mar. 1943, *Steyermark 52325 a* [F!]. **Dpto. Retalhuleu:** Along Río Samalá, between San Sebastián and Santa Cruz Muluá, (14° 36' N, 91° 36' W), 330 m, 23 Feb. 1941, *Standley 88162* [F!, GH!]. **Dpto. Suchitepequez:** Between Cocales and San Lucas Toliman, in Santa Teresa, Apr. 1990, (14° 25' N, 91° 05' W), *Welz 3124* [B!, FR!, HBG!, LPB!, SEL!, U!]; BGB!, FRP!, BGHD!, BGHH!, BGU!]; ibid.: Finca Moca, 1000 m, (14° 30' N, 91° 30' W), 4 Jan. 1935, *Skutch 2064* [F!, GH!, US!, NY!]; ibid.: Along Río Madre Vieja above Patulul, (14° 25' N, 91° 10' W), 450 m, 6 Jan. 1939, *Standley 62215* [F!]. **Dpto. Escuintla:** Along Río Michatoya, SE of Esquintla, (14° 12' N, 90° 42' W), 300 m, 12 Mar. 1941, *Standley 89029* [F!]; ibid.: 12 km W of Escuintla, (14° 18' N, 90° 54' W), 300 m, 9 Nov. 1970, *Harmon & Fuentes 4682* [MO!, US!]; ibid.: Along Río Guacalate, (14° 24' N, 90° 49' W), 520 m, 28 Nov. 1938, *Standley 58282* [F!, GH!]; ibid.: Along Río Guacalate, (14° 24' N, 90° 49' W), 600 m, 16 Dec. 1938, *Standley 60180* [F!, GH!]; ibid.: Next to Esquintla, 5 Dec. 1938, (14° 18' N, 90° 48' W), 5 Dec. 1938, *Johnston 1521* [F!]; ibid.: San Antonio Jute, (14° 18' N, 90° 48' W), 780 m, 9 Feb. 1939, *Standley 64890* [F!]; ibid.: 46 km on route CA 9 from Guatemala-City to Esquintla, (14° 18' S, 90° 42' W), 650 m, 3 Nov. 1993, *Radtke & Moldenhauer 212* [B!; BGB!]. **Dpto. Sacatepequez:** Las Lajas, (14° 30' N, 90° 45' W), 1200 m, 28 Nov. 1938, *Standley 58297* [F!]; ibid.: Below Barranco Hondo (14° 30' N, 90° 45' W), 1100m, 11 Mar. 1941, *Standley 89001* [F!].
EL SALVADOR: **Dpto. Ahuachapán:** San Benito, W of Piedra del Mancho, mountain Tacho Lopez, 13° 49' N, 89° 56' W, 19 Nov. 1992, *Sandoval 837* [LAGU (phot.!), MO!]; ibid.: Near Salto de Atehuecía, (13° 47' N, 89° 55' W), 600 m, 22 Jan. 1947, *Standley 2881* [F!, US!]; ibid.: Paso Miguel López, 13° 49' N, 89° 56' W, 450 m, 19 Dec. 1997, *Sandoval 1717* [LAGU (phot.!), MO! WU!]; ibid.: San Francisco Menéndez, El Corozo, Mariposario, zona baja "Mariposario", 13° 49' N, 89° 59' W, 200 m, *Rosales 1821* [LAGU (phot.!)]; ibid.: San Francisco Menéndez, El Corozo, Mariposario, 13° 49' N, 89° 59' W, 200 m, 9 Nov. 2000, *Rosales 1818* [LAGU (phot.!), MO!]; ibid.: Parque Nacional El Imposible, Río Guayapa, 13° 49' N, 89° 56' W, 850 m, 12 Nov. 1991, *Villacorta 932* [LAGU (phot.!), MO!]; ibid.: Parque Nacional El Imposible, Río Las Positas, 13° 51' 06" N, 89° 59' 12" W, 500 m, 18 Jan. 1998, *Monro 1913* [LAGU (phot.!), MO!]; ibid.: Parque Nacional El Imposible, Pacayal, 13° 52' N, 89° 59' W, 400 m, 11 Dec. 1990, *Berendsohn 1325* [LAGU (phot.!)]; ibid.: San Francisco Menéndez, Parque Nacional El Imposible, Hda. San Benito, union of Río Guayapa and Río Venado, 13° 49' N, 89° 56' W, 500 m, 23 Nov. 1991, *Berendsohn 1420* [LAGU (phot.!)]. **Dpto. Chalatenango:** Track from Hacienda Santa Cruz to Laguna, SW of Carizal, (14° 02' N, 88°56' W), 800 m, 25 Nov. 1950, *Rohweder 1* [HBG!, MO!]. **Dpto. La Libertad:** National Park Deininger, 13° 30' N, 89° 18' W, 200 m, *Welz s.n.*

[B!, FR!, LPB!; BGB!]; ibid.: 28 km from Playa El Majanual, 13° 30' 23" N, 89° 30' 22" W, 130 m, 29 Jan. 1998, *Sidwell 467* [BM!, MO!]; ibid.: 13° 14' N, 89° 15' W, 800 m, 14 Dec. 1988, *Villacorta 221* [LAGU (phot.!), SEL (phot.!)]. **Dpto. Cuscatlán:** Tenancingo, Canton Huiliguiste, Quebrada la Tigra, 13°50'50" N, 88°56'55" W, 425 m, 16 Nov. 2007, *Menjívar 1780* [MHES (phot.!)]; ibid.: Tenancingo, Canton Huiliguiste, Quebrada la Tigra, 13°50'50" N, 88°56'55" W, 425 m, 16 Nov. 2007, *Morales 15882* [MHES (phot.!)]. **Dpto. Cabañas:** Cinquera, Zona Protegida, Río San Benito, 374 m, 13° 53' N, 88° 57' W, 23 Jan. 2003, *Carballo 600* [LAGU (phot.!), MO!].
WITHOUT LOCALITY: *s.n.* [B 37883!; BGB 185-12-37-80!; ex HBV]; *s.n.* [B 17279!; BGB 136-10-79-80/83! ex HBV].

Distribution and ecology

Range size:	Relatively large.
Countries:	MEXICO. Dpto. Guerrero, Veracruz, Oaxaca, Chiapas. GUATEMALA. Dpto. Huehuetenango, San Marcos, Sacatepequez, Retalhuleo, Suchitepequez, Escuintla. EL SALVADOR. Dpto. Chalatenango, La Libertad, Ahuachapan, Cuscatlán, Cabañas.
Ecoregions:	Sinaloan-Dry-Forests, Jalisco-Dry-Forests, Veracruz-Moist-Forests, Petén-Verycruz-Moist-Forests, Central-American-Pine-Oak-Forests, Southern-Pacific-Dry-Forests, Central-American-Dry-Forests, Sierra-Madre-de-Chiapas-Moist-Forests.
Life style & habitats:	More or less dry tropical to subtropical forests. Terrestrial or saxicolous, on moist, shady rocks, steep slopes, riverbanks, roadsides and small ravines. Very common.
Altitude:	100–1300 m.

Taxonomic delimitation and systematic relationships

Fosterella micrantha is morphologically very similar to *F. villosula* and molecular data indicate a very close relationship as well. Together with *F. christophii*, which bears striking resemblance as well, they form a distinct, highly supported clade within the genus.

Fosterella micrantha differs from *F. villosula* and *F. christophii* by the light green, never reddish leaf blades, only slightly villous inflorescence axes and floral bracts and glabrous sepals. Furthermore, they differ ecologically, as the widely distributed *F. micrantha* from Mexico, Guatemala and El Salvador is distributed in more or less dry tropical to subtropical forests, whereas *F. villosula* and *F. christophii* are both characterised by relatively small distributional ranges in the Bolivian Andes.

Vernacular name and domestic use

Fosterella micrantha is called "Cola de gallo" in El Salvador (*Berendsohn 1420* [LAGU!]). In Mexico squashed leaves mixed with cold water are used as remedy against fever (ESPEJO-SERNA et al. 2005).

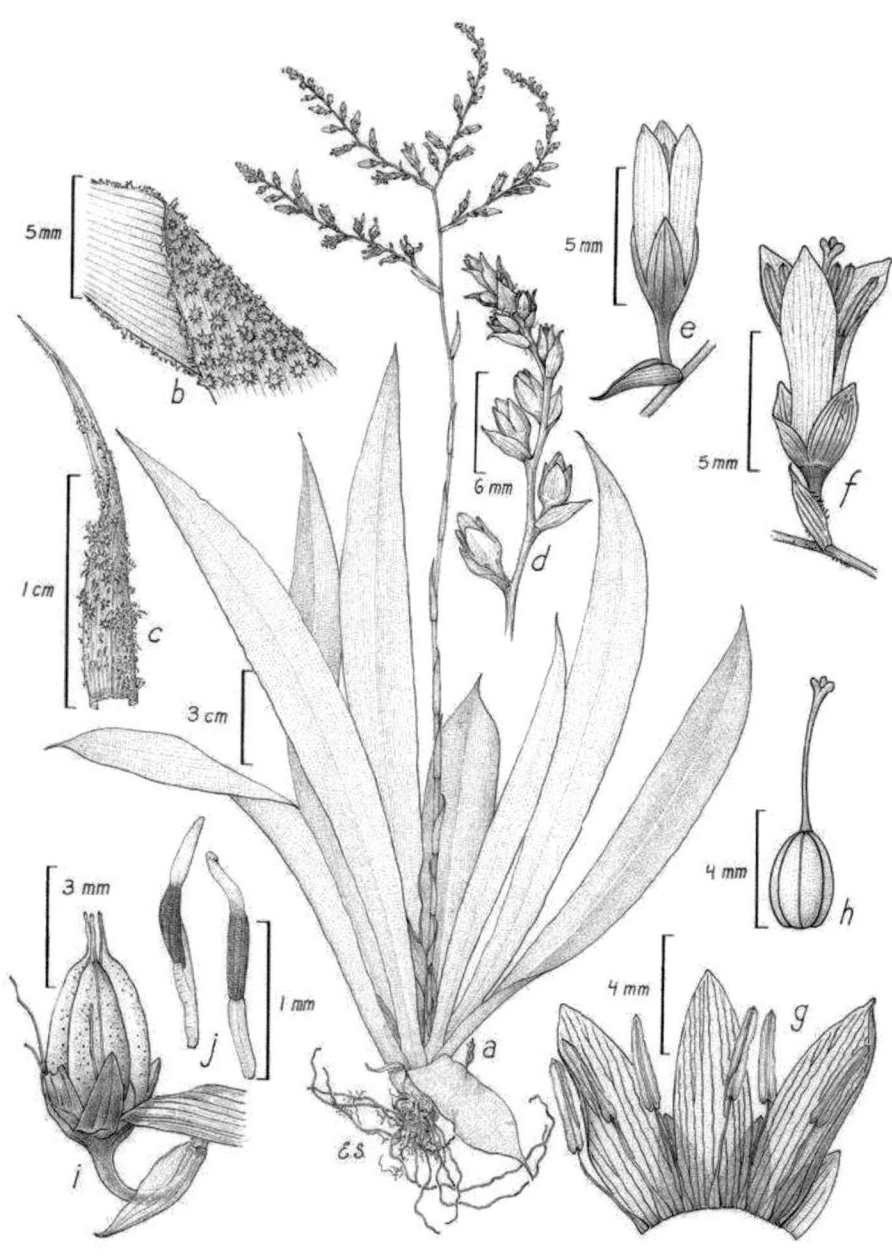

Fig. 68: *Fosterella micrantha*. A Habit. B Indumentum of the abaxial leaf surface. C Peduncle bract. D Inflorescence branch. E Flower before anthesis. F Flower during anthesis. G Petals and stamens. H Gynoecium. I Fruit. J Seeds. Drawing of *Vásquez 223* by E. Saavedra in Espejo-Serna et al. (2005). © Flora de Veracruz.

Fig. 69: Holotype of *Fosterella micrantha* (Lindl.) L.B. Sm. [CGE].

Fig. 70: *F. micrantha (Welz 3124).*

Fig. 71: Flowers of *F. micrantha (Welz 3124).*

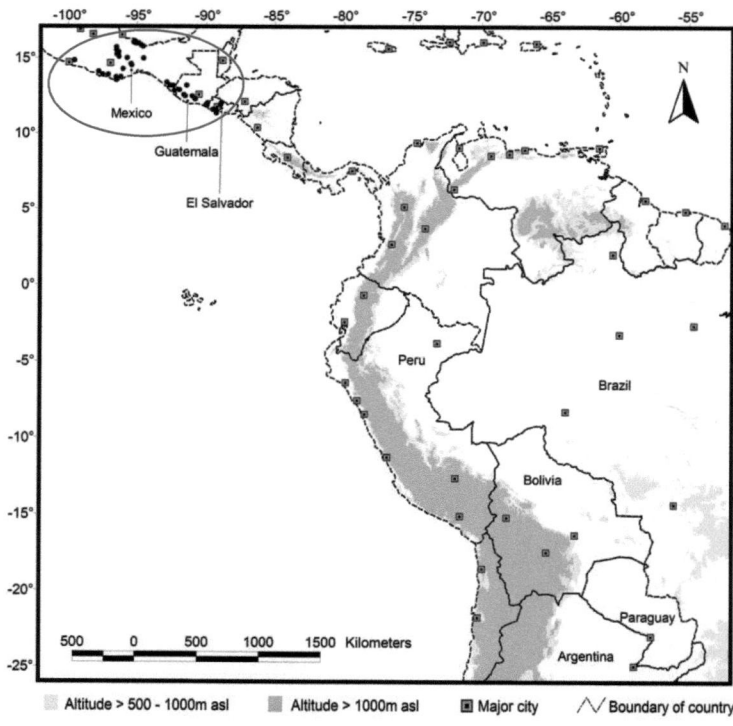

Fig. 72: Distribution of *Fosterella micrantha.*

Fosterella nicoliana J. Peters & Ibisch, Die Bromelie 2008 (2): 66. 2008. Fig. 73–75.

> TYPE: Perú. Dpto. San Martín/Loreto: Vicinity of Aguaytía, woods west of chacra of Don Diogenes del Aguila, *M.E. Mathias & D. Taylor 5988* [Holotype: F!]. Perú. Dpto. San Martín/Loreto: Vicinity of Aguaytía, high ground in forest southeast of house, Don Diogenes del Aguila, east of Aguaytía, between Pucallpa Road and Río Aguaytía, 29 June 1960, *M.E. Mathias & D. Taylor 5042* [Paratype: F!, MO!].

Comment on type

The Vicinity of Aguaytía is situated in the Dpto. Ucayali, approx. at 09° 02' S, 75° 30' W.

Etymology

The taxon is dedicated to the German botanist Nicole Schütz (*1977) who is currently promoting research on Bromeliaceae by preparing a monograph of the genus *Deuterocohnia* Mez.

Description

Plant acaulescent, up to 40 cm high. **Leaves** few, up to 12, forming an open, more or less upright rosette. **Sheaths** 10–20 mm wide, entire, whitish, glabrous. **Blades** lanceolate, acuminate, narrowed towards the base, to 40 cm long and 5 cm wide, thin, entire, adaxially glabrous, abaxially scattered lepidote by peltate trichomes with dentate margin, glabrescent. **Peduncle** 15–25 cm long, 2 mm in diameter, arachnoid. **Peduncle bracts** 1.5 cm long, equalling the internodes, entire, abaxially arachnoid. **Inflorescence** compound racemose, to 15 cm long and 10 cm wide, axes green, arachnoid. **Primary bracts** 6–8 mm long, equalling the sterile base of the branches, entire, abaxially arachnoid. **Branches** 5–8, inclined, arcuate, to 7 cm long, bearing up to 20 flowers. **Floral bracts** 1–2 mm long, equalling the pedicels, entire, abaxially sparsely arachnoid. **Flowers** secund, pendulous, 3–5 mm apart. **Pedicels** to 2 mm long. **Sepals** 2–3 mm long, green, glabrous. **Petals** whitish, 5–6 mm long, slightly recurved during anthesis, straight afterwards. **Filaments** 3 mm long. **Anthers** 2 mm long. **Style** 2 mm long. **Stigmatic complex** simple-erect. **Capsule** globose, 4 mm long, 4 mm wide. **Seeds** clavate.

Specimens seen

PERÚ. **Dpto. Pasco:** Prov. Oxapampa Pichis Valley, Near Paujil, 10 km downriver from Puerto Bermúdez, E side of river across from big bend with large island, 74° 53' W, 10° 15' S, 300 m, 23 Sept. 1982, *Foster 8868* [F!, MO!].

Distribution and ecology

Range size:	Very small.
Countries:	PERÚ. Dpto. Ucayali, Pasco.
Ecoregion:	Ucayali-Moist-Forest.
Life style & habitats:	Humid, evergreen tropical rain forest. Forest on low hills, riverbanks and small ravines (*Foster 8868* [MO]).
Altitude:	300 m.

Taxonomic delimitation and systematic relationships

Fosterella nicoliana differs from all other species of the genus by its globose capsules and clavate seeds, which show no differentiation in cell-patterns of the seed-surface: both the seed itself and the appendages exhibit a reticulate and isoquadrangular structure. In all other *Fosterella* species, the seeds comply with the "*Fosterella*-seed-type" described by VARADARAJAN & GILMARTIN (1988 a): filiform seeds, carrying bicaudate appendages, characterised by cell-patterns, that differ between the surface of the seeds itself and that of the appendages: reticulate and isoquadrangular on the seed surface and elongated on the surface of the appendages. Thus, this new type of seeds within the genus is even distinct from the "*Fosterella*-seed-type" described by VARADARAJAN & GILMARTIN (1988 a), supposed to be typical for *Abromeitiella, Ayensua, Cottendorfia, Deuterocohnia, Fosterella, Hechtia* and *Pitairnia*. So far, this species has only been known from herbarium vouchers and no molecular data is yet available.

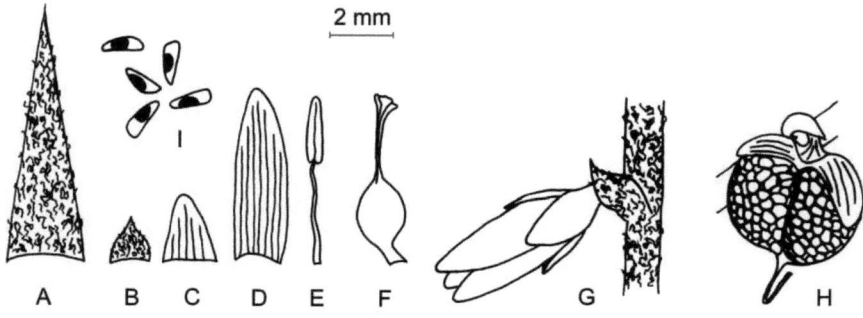

Fig. 73: *Fosterella nicoliana*. A Primary bract. B Floral bract. C Sepal. D Petal. E Stamen. F Gynoecium. G Flower. H Fruit. I Seeds. Drawing of *Mathias & Taylor 5988* by J. Peters in PETERS et al. (2008 a). © Die Bromelie.

Fig. 74: Holotype of *Fosterella nicoJana* J. Peters & Ibisch [F].

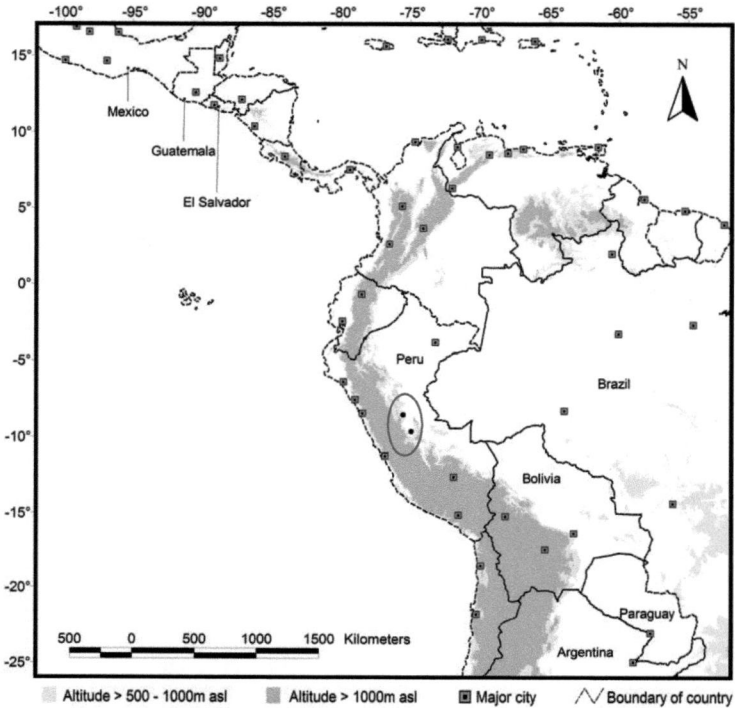

Fig. 75: Distribution of *Fosterella nicoliana*.

Fosterella pearcei (Baker) L.B. Sm., Phytologia 7: 172. 1960. Fig. 76 & 77.

Basionym:	*Cottendorfia pearcei* Baker, Handbook of the Bromeliaceae: 128. 1889. TYPE: Western slopes of the Andes, Butuco, July 1865, *Pearce s.n.* [Holotype: BM (phot.!)].
≡	*Lindmania pearcei* (Baker) Mez in C.DC., Monogr. Phan. 9: 537. 1896.

Comment on type

BAKER (1889), in describing *Cottendorfia pearcei*, gave the type locality as "Butuco" which is a misinterpretation of Pearce's handwriting which said "Buturo" as was already assumed by SMITH & DOWNS (1974). In addition, it was stated that the collection came from the "Western slopes of the Andes". This led MEZ (1896) to the erroneous conclusion that Colombia is the home country of the species. In fact, the label gives no other information than "Buturo", and from the data on other specimens collected by Pearce the conclusion must be drawn that it is located in Bolivia. Buturo definitely is a Bolivian village on the eastern slopes of the Andes and is located close to Asariamas in the Province of Abel Iturralde in the Department of La Paz (14° 16' S, 68° 34' W, approx. 990 m). When BAKER (1889) described *Cottendorfia pearcei*, he listed just the single specimen *Pearce s.n.* (BM). Therefore, SMITH & DOWNS (1974) were correct in citing it as holotype.

Etymology

The species is dedicated to Richard William Pearce (ca. 1835–1868), British plant collector in South America for the nursery John Gould Veitch and collector of the type specimen.

Description

Plant acaulescent, up to 50 cm high. **Leaves** few, up to 12, forming an open, more or less upright rosette. **Sheaths** 30–40 mm wide, entire, whitish, glabrous. **Blades** linear, acuminate, narrowed towards the base, to 45 cm long and 1–1.5 cm wide, thin, entire, adaxially glabrous, abaxially densely appressed lepidote by peltate trichomes with fimbriate margin. **Peduncle** to 80 cm long, to 5 mm in diameter, green, densely arachnoid. **Peduncle bracts** 6–12 cm long, much longer than the internodes, entire, abaxially appressed lepidote. **Inflorescence** paniculate with branches up to 2nd order, 45 cm long and 15 cm wide, axes green, arachnoid. **Primary bracts** to 10 mm long, slightly shorter than the sterile base of the branches, entire, abaxially appressed lepidote. **Branches** up to 10, inclined, arcuate, 3–10 cm long, bearing up to 20 flowers. **Secondary branches** 2–5 cm long, bearing up to 10 flowers. **Floral bracts** 3–4 mm long, longer than the pedicels, entire, abaxially appressed lepidote, glabrescent. **Flowers** spreading, suberect, 3–7 mm apart. **Pedicels** to 1.5 mm long. **Sepals** 2–3 mm long, green, glabrous. **Petals** 5 mm long, white, recoiled like watchsprings during anthesis and afterwards. **Filaments** 3 mm long. **Anthers** 2 mm long. **Style** 2 mm long. **Stigmatic complex** simple-erect. **Capsule** ovoid, 5 mm long, 3 mm wide. **Seeds** not seen.

Specimens seen

BOLIVIA: **Dpto. La Paz:** Prov. Iturralde: Parque Nacional Madidi, Mamacona, 38 km E of Apolo, track towards San José de Uchupiamonas, 14° 27' 33" S, 68° 11' 05" W, 1700–2100 m, 10 July 2002, *Fuentes 4893* [MO!]. Prov. Sud Yungas: Valley of Río Boopi, San Bartolome, near Calisaya, (16° 12' S, 67° 11' W), 800 m, July 1939, *Krukoff 10328* [F!, GH!, K!, LPB!, NY!, U!].
WITHOUT LOCALITY: *s.n.* [SEL 090018!].

Distribution and ecology

Range size: Very small.

Country: BOLIVIA. Dpto. La Paz.

Ecoregion: Yungas.

Life style & habitats: Humid, evergreen montane forests, in secondary forests and shrubberies, along shady roadsides.

In humid montane forest with *Podocarpus oleifolius* (Podocarpaceae), *Alchornea* sp. (Euphorbiaceae), Lauraceae and Melastomataceae, secondary shrubberies caused by fire with *Chusquea* sp. (Poaceae), Melastomataceae, Asteraceae, *Pteridium arachnoideum* (Dennstaedtiaceae) and Ericaceae (*Fuentes 4893* [MO]).

Altitude: 800–2000 m.

Taxonomic delimitation and systematic relationships

Fosterella pearcei is similar to *F. albicans*, but differs in having less leaves which are also shorter and narrower and show less pronounced indumentum on the abaxial leaf surface. The primary bracts are shorter and the inflorescence is less branched. Hitherto no molecular data is on hand for *F. pearcei*.

Fig. 76: Holotype of *Fosterella pearcei* (Baker) L.B. Sm. [BM]. © The Natural History Museum, London.

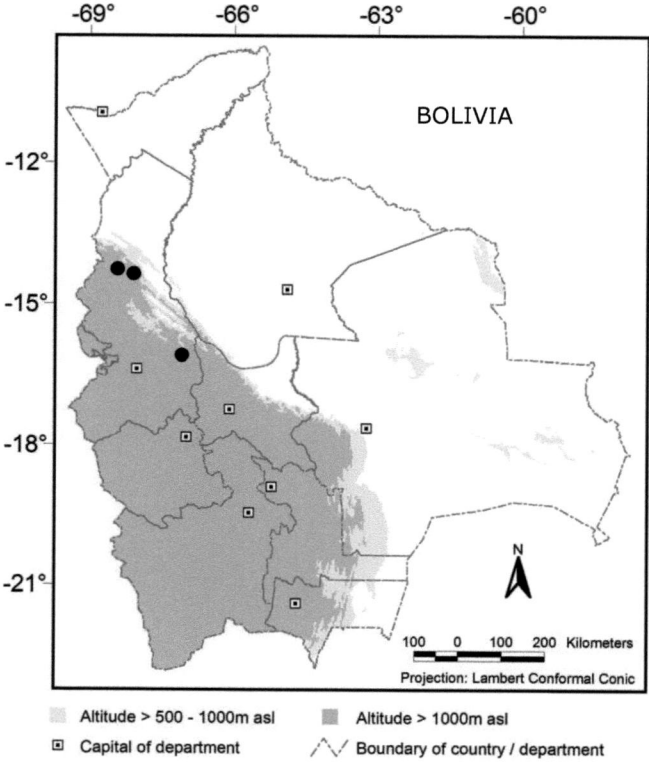

Fig. 77: Distribution of *Fosterella pearcei*.

Fosterella penduliflora (C.H. Wright) L.B. Sm., Phytologia 7: 172. 1960. Fig. 78–82.

Basionym:	*Catopsis penduliflora* C.H. Wright, Bull. Misc. Inform. Kew: 197. 1910. TYPE: Perú, without precise locality, *M. Forget s.n.*, described from a living plant sent by Messrs. F. Sander & Sons, 1910 [Lectotype: K!].
≡	*Lindmania penduliflora* (C.H. Wright) Stapf, Bot. Mag. 150: Pl. 9029. 1924.
=	*Fosterella chiquitana* Ibisch, R. Vásquez & E. Gross, Rev. Soc. Boliviana Bot. 2 (2): 118. 1999. TYPE: Bolivia. Dpto. Santa Cruz: Prov. Ñuflo de Chávez, about 10 km from Concepción on the road to San Javier, on granitic outcrops (zonal vegetation semideciduous-semihumid Chiquitano forest), 16° 10' S, 62° 05' W, 500 m, 16 May 1997, *P. Ibisch 98.0125*, flowered in the garden of Pierre L. Ibisch, Santa Cruz de la Sierra, from 21 Aug. 1998 onwards [Holotype: LPB!; Isotypes: FR!, WU!].
=	*Fosterella latifolia* Ibisch, R. Vásquez & E. Gross, Rev. Soc. Boliviana Bot. 2 (2): 123. 1999. TYPE: Bolivia. Dpto. Santa Cruz: Prov. Florida, Refugio Volcanes, in understory of semihumid forests, 18° 05' S, 63° 40' W, 1000 m, 16 May 1998, *P. Ibisch 98.0098* [Holotype: LPB!; Isotype: SEL!].

Comment on type

C.H. WRIGHT (1910), in describing *Catopsis penduliflora* from a living plant sent by Messrs. F. Sanders & Sons, did not cite a voucher nor chose a holotype. STAPF (1924), who recombined this species to *Lindmania penduliflora*, prepared detailed drawings on a herbarium specimen preserved at Kew in 1923. Obviously, he studied a voucher of the same living plant, which Wright had described and which was still cultivated in Kew at that time. Consequently, we designate that specimen as lectotype (K000321535).

Clonotypes of *Fosterella chiquitana* Ibisch, R. Vásquez & E. Gross are cultivated in the Botanical Gardens Heidelberg (Germany) and in the living collection of the Fundación Amigos de la Naturaleza, Santa Cruz (Bolivia).

A Clonotype of *Fosterella latifolia* Ibisch, R. Vásquez & E. Gross is cultivated in the living collection of the Fundación Amigos de la Naturaleza, Santa Cruz (Bolivia).

Etymology

The epithet refers to the pendulous flowers.

Description

Plant acaulescent, up to 60 (120) cm high. **Leaves** few, 8–15, forming an open, flat rosette. **Sheaths** 20–50 mm wide, entire, whitish, glabrous. **Blades** lanceolate, acuminate, narrowed towards the base, 20–40 cm long and 2.5–8 cm wide, succulent towards the base, entire, adaxially glabrous, abaxially frequently reddish, scattered lepidote by peltate trichomes with dentate margin. **Peduncle** 25–60 (85) cm long, 2–5 (7) mm in diameter, green, frequently reddish, glabrous, frequently glaucous. **Peduncle bracts** 1–3 cm long, equalling the internodes, entire, frequently reddish, abaxially scattered lepidote. **Inflorescence** paniculate with branches up to 2nd order, 10–40 cm long and 10–25 cm wide, axes frequently reddish or glaucous, glabrous. **Primary bracts** 5–15 mm long, shorter than the sterile base of the branches, entire, frequently reddish, abaxially scattered lepidote. **Branches** 10–15, inclined, arcuate, 8–22 cm long, bearing up to 35 flowers. **Secondary branches** 3–5 cm long, bearing 5–10 flowers. **Floral bracts** 2–4 (10) mm, equalling the pedicels, entire, frequently reddish, abaxially scattered lepidote to villous, glabrescent. **Flowers** secund, pendulous, 5–10 mm apart. **Pedicels** 2–6 mm long. **Sepals** 2–3 mm long, green, frequently reddish, glabrous. **Petals** 7–10 mm long, white, recurved during anthesis, straight afterwards. **Filaments** 4–7 mm long. **Anthers** 2–3 mm long. **Style** 3–5 mm long. **Stigmatic complex** conduplicate-spiral. **Capsule** ovoid, 5–7 mm long, 3–4 mm wide. **Seeds** filiform, bicaudate, 2 mm long.

Specimens seen

PERÚ: Without precise locality, *Goebel s.n.* [FR!, HEID!; BGHD!].
BOLIVIA: Without precise locality, *Vásquez 4801* [VASQ!]; Without precise locality, July 1979, *Hromadnik 5225* [HEID!]. **Dpto. La Paz:** Prov. Franz Tamayo: Parque Nacional Madidi, track from Apolo to Azariamas, Arroyo Pintata, 14° 27' 54" S, 68° 32' 56" W, 938 m, 3 Mar. 2003, *Canqui 252* [LPB! (st.), WU! (st.)]. Prov. Nor Yungas: 25 km from Caranavi to Yolosa, 15° 57' S, 67° 34' W, 830 m, 15 Feb. 2000, *Vásquez & Gerlach 3674* [FR!, SEL!; BGHD!]. Prov. Sud Yungas: 10 km from Irupana to Chulumani, 16° 27' S, 67° 28' W, 1400 m, 12 Feb. 2000, *Vásquez 3636 a1* [FR!, LPB!, VASQ!; FAN!]; ibid.: 84 km from Coroico to Chulumani, 2650 m, 1997, *Gouda s.n.* [B!, FR!, LPB!, WU!; BGB!, FRP!, BGU!, HBV!]. Prov. Inquisivi: Licoma Cruce between Circuata and Cajuata, 16° 44' S, 67° 13' W, 2430 m, 11 Feb. 2000, *Vásquez 3624* [FR!; FAN!]. **Dpto. Cochabamba:** Without precise locality, 1200 m, 22 Apr. 1985, *Cathcart s.n.* [HEID!, WU!; BGDD!, BGOS!, HBV!]. **Dpto. Santa Cruz:** Prov. Guarayos: Along the road between Trinidad und Santa Cruz, 2 km before Santa Maria, 15° 40' 06" S, 63° 31' 37" W, 200 m, 30 Oct. 2002, *Rex & Schulte 301002-1* [FR!]; ibid.: Along the road between Trinidad und Santa Cruz, 2 km before Santa Maria, 15° 40' 12" S, 63° 31' 37" W, 200 m, 30 Oct. 2002, *Rex & Schulte 301002-2* [LPB!]; ibid. Along the road between Trinidad und Santa Cruz, 2 km before Santa Maria, 15° 40' 06" S, 63° 31' 37" W, 200 m, 30 Oct. 2002, *Rex & Schulte 301002-3* [FR!, VASQ!]; ibid.: Between San Pablo and Ascensión de Guarayas, Cerro Grande, 15° 38' S, 63° 30' W, 220 m, 29 Oct. 2002, *Vásquez 4685* [VASQ!; FAN!]. Prov. Ñuflo de Chavez: Between San Javier and Concepción, 16° 14' S, 62°13' W, 30 July 2000, *Vásquez 3762* [FR!, HEID!, LPB!, SEL!, VASQ!; BGHD!]; ibid.: About 10 km from Concepcion to San Javier, 16° 10' S, 62° 05' W, 500 m, 16 May 1997, *Ibisch & Ibisch 98.0125* [FR!, HEID!, LPB!, SEL!, WU!; FAN!, BGHD!]; ibid.: Between San Javier and Concepción, 16° 14' S, 62° 13' W, 22 Apr. 2000, *Nowicki 2221* [FR!, LPB!, SEL!, USZ!; BGHD!]. Prov. Florida: Refugio Los Volcanes, between Santa Cruz and Samaipata, 3 km NE of Bermejo, 18° 06' S, 63° 36' W, 1050 m, 3 Oct. 1997, *Kessler 12273* [SEL!]; ibid.: Refugio Los Volcanes, between Santa Cruz and Samaipata, NE of Bermejo, property of Albert Schwiening, track towards El Sillar, 18° 05' S, 63° 40' W, 1000 m, 16 May 1995, *Ibisch 98.0098* [FR!, LPB!, SEL!, VASQ!, WU!]; ibid.: Between La Angustura y Bermejo, 18° 10' S, 63° 35' W, 700 m, 13 Oct.

1999, *Vásquez 3406* [FR!, LPB!, SEL!, USZ!, VASQ!, WU!]; ibid.: Between La Angostura y Bermejo, 18° 10' S, 63° 35' W, 700 m, 13 Oct. 1999, *Vásquez & Quispe 3407* [FR!, LPB!, SEL!, VASQ!; FAN!]; ibid.: Between Cuevas and Bella Vista, 18° 13' S, 63° 44' W, 1350 m, 10 Sept. 1999, *Vásquez 2979* [FR!, HEID!, LPB!, SEL!, USZ!, WU!; FAN!]; ibid.: Bermejo, 87 km from Santa Cruz to Cochabamba, 18° 12' S, 63° 36' W, 5 May 1992, *Seddi 250* [USZ!]; ibid.: Between Samaipata and Bermejo, 18° 08' 24" S, 63° 31' 37" W, 1019 m, 11 Oct. 2002, *Rex & Schulte 111002-1* [FR!; FAN!]; ibid.: Between Samaipata and Bermejo, 18° 08' 24" S, 63° 31' 37" W, 1019 m, 11 Oct. 2002, *Rex & Schulte 111002-2* [SEL!]; ibid.: Between Samaipata and Bermejo, 18° 08' 24" S, 63° 31' 37" W, 1019 m, 11 Oct. 2002, *Rex & Schulte 111002-3* [LPB!]; ibid.: Between Samaipata and Bermejo, 18° 08' 24" S, 63° 31' 37" W, 1019 m, 11 Oct. 2002, *Rex & Schulte 111002-4* [FR!; FAN!]; ibid.: Between Samaipata and Bermejo, 18° 08' 24" S, 63° 31' 37" W, 1019 m, 11 Oct. 2002, *Rex & Schulte 111002-5* [LPB!; FAN!]; ibid.: Between Samaipata and Bermejo, 18° 08' 24" S, 63° 31' 37" W, 1019 m, 11 Oct. 2002, *Rex & Schulte 111002-6* [SEL!; FAN!]; ibid.: Between Samaipata and Bermejo, 18° 08' 24" S, 63° 31' 37" W, 1019 m, 11 Oct. 2002, *Rex & Schulte 111002-7* [FR!; FAN!]; ibid.: Between Samaipata and Bermejo, 18° 08' 24" S, 63° 31' 37" W, 1019 m, 11 Oct. 2002, *Rex & Schulte 111002-8* [LPB!; FAN]; ibid.: Between Samaipata and Bermejo, 18° 08' 24" S, 63° 31' 37" W, 1019 m, 11 Oct. 2002, *Rex & Schulte 111002-9* [SEL!]; ibid.: Between Samaipata and Bermejo, 18° 08' 24" S, 63° 31' 37" W, 1019 m, 11 Oct. 2002, *Rex & Schulte 111002-10* [LPB!; FAN!]. Prov. Andres Ibañez: Road from Bermejo to Angosturo, 18° 10' 42" S, 63° 34' 30" W, 767 m, 27 Oct. 2006, *Peters 06.0112* [LPB!; BGHD!]; ibid.: El Torno, 17° 54' S, 63° 24' W, 2002, *Ibisch 02.0006* [LPB!, SEL!, WU!; BGHD!]. Prov. Chiquitos: San José de Chiquitos, 17° 48' S, 60° 42' W, June 2003, *Reichle P-SR 2* [FR!, LPB!, SEL!, WU!; BGHD!]. Prov. Vallegrande: 10 km from Masicuri towards Valle Grande, 18° 47' S, 63° 50' W, 1000 m, 14 July 1995, *Kessler 5373* [SEL!]. Prov. Cordillera: Alto Parapetí, Hacienda Yapuimbia, 800 m, 30. Sept. 1985, *Michel 435* [LPB!, MO!]; ibid.: 33 km from Ipati to Lagunillas, 19° 48' 18" S, 63° 42' 37 ' W, 1280 m, 3 Oct. 2006, *Peters 06.0018* [LPB!; BGHD!]; ibid.: 33 km from Ipati to Lagunillas, 19° 48' 18" S, 63° 42' 37" W, 1280 m, 3 Oct. 2006, *Peters 06.0019* [SEL!; BGHD!]; ibid.: 45 km from Ipati to Lagunillas, 19° 51' 14" S, 63° 43' 50" W, 1416 m, 3 Oct. 2006, *Peters 06.0022* [FR!]; ibid.: Lagunillas, Cordillera Caro Huayaco, (19° 39' S, 63° 40' W), 1200 m, Aug. 1934, *Cardenas 2856 a* [F!, GH!]; ibid.: 8 km W of Ababó, bridge across Río Grande, 18° 54' S, 63° 24' W, 21 Apr. 2000, *Ibisch & Vásquez 00.0036* [FR!, LPB!; FAN!]; ibid.: 16 km from Ababó towards Gutierrez, (18° 54' S, 63° 24' W), 650 m, 22 Dec. 2000, *Vásquez 3817* [FR!, HEID!, LPB!, SEL!, USZ! WU!; FAN!, BGHD!]; ibid.: 13 km from Cruce de Lagunillas to Incahuasi, entry to the canyon of Incahuasi, 19° 49' S, 63° 33' W, 1120 m, 22 Apr. 2000, *Vásquez & Vásquez 3724 a* [FR!, LPB!, SEL!, VASQ!, WU!; FAN!]; ibid.: 13 km from Cruce de Lagunillas to Incahuasi, entry to the canyon of Incahuasi, 19° 49' S, 63° 33' W, 1120 m, 22 Apr. 2000, *Vásquez 3724 b* [LPB!, VASQ!]; ibid.: 67 km from Camiri towards Monteagudo, (19° 56' S, 63° 33' W), 1150 m, 27 Oct. 1983, *Beck & Liberman 9781* [LPB!, SEL!, USZ!]. **Dpto. Chuquisaca:** Road between Monteagudo and Sucre, Aug. 1993, *Cathcart B-46* [SEL!; MSBG]. Prov. B. Boeto: Villa Serrano, between Pampas del Tigre and Temporal, 19° 04' 47" S, 64° 10' 32" W, 1529 m, 25 Nov. 2005, *Villalobos & Paredes 222* [HSB!]. Prov. Jaime Mencoza: Next to Llantoj, 19° 19' 00" S, 64° 04' 08" W, 1300 m, 11 Oct. 2004, *Gutiérrez 994* [HSB!, LPB!]; ibid.: 68 km from Monteagudo to Padilla; 19° 34' S, 64° 09' W, 1400 m, *Kessler 4987* [SEL! (st.)]. Prov. Hernando Siles: 48 km from Monteagudo to Padilla, 19° 38' S, 64° 03' W, 1100 m, 28 June 1995, *Kessler 4888* [SEL! (st.)]; ibid.: Between Padilla and Monteagudo, Puente Azero, (19° 42' S, 64° 06' W), 1400 m, 19 Nov. 1994, *Wood 8843* [K!, LPB!]; ibid.: 9 km from Monteagudo to Padilla, 19° 48' 11" S, 64° 01' 15" W, 1153 m, 3 Oct. 2006, *Peters 06.0026* [LPB!]; ibid.: 9 km from Monteagudo to Padilla, 19° 48' 11" S, 64° 01' 15" W, 1153 m, 3 Oct. 2006, *Peters 06.0027* [SEL!]; ibid.: 9 km from Monteagudo to Padilla, 19° 48' 11" S, 64° 01' 15" W, 1153 m, 3 Oct. 2006, *Peters 06.0028* [FR!]; ibid.: Road from Monteagudo to Padilla, next to Monteagudo, 19° 47' 13" S, 64° 02' 23" W, 1216 m, 3 Oct. 2006, *Peters 06.0035* [LPB!; BGHD!]; ibid.: Road from Monteagudo to Padilla, next to Monteagudo, 19° 47' 13" S, 64° 02' 23" W, 1216 m, 3 Oct. 2006, *Peters 06.0036* [SEL!; BGHD!]; ibid.: 34 km from Monteagudo to Padilla, next to Bartolo, 19° 39' 52" S, 64° 02' 49" W, 1346 m, 4 Oct. 2006, *Peters 06.0038* [FR!; BGHD!]; ibid.: 34 km from Monteagudo to Padilla, next to Bartolo, 19° 39' 52" S, 64° 02' 49" W, 1346 m, 4 Oct. 2006, *Peters 06.0039* [BGHD!]; ibid.: 34 km from Monteagudo to Padilla, next to Bartolo, 19° 39' 52" S, 64° 02' 49" W, 1346 m, 4 Oct. 2006, *Peters 06.0040* [BGHD!]; ibid.: 34 km from Monteagudo to Padilla, next to Bartolo, 19° 39' 52" S, 64° 02' 49" W, 1346 m, 4 Oct. 2006, *Peters 06.0041* [BGHD!]; ibid.: 50 km from Monteagudo to Padilla, 19° 36' 55" S, 64° 03' 49" W, 1147 m, 4 Oct. 2006, *Peters 06.0042* [LPB!; BGHD!]; ibid.: 50 km from Monteagudo to Padilla, 19° 36' 55" S, 64° 03' 49" W, 1147 m, 4 Oct. 2006, *Peters 06.0043* [SEL!]; ibid.: 64 km from Monteagudo to Padilla,

19° 32' 19" S, 64° 07' 28" W, 1269 m, 4 Oct. 2006, *Peters 06.0050* [LPB!]; ibid.: 40 km from Monteagudo to Padilla, 19° 40' S, 64° 03' W, 1300 m, 29 June 1995, *Kessler 4848 b* [SEL!]; ibid.: Pirairimi, (18° 38' S, 63° 58' W), 2640 m, 18 Nov. 1988, *Murguia & Munoz 193* [LPB!, SEL!]. Prov. Luis Calvo: Between Incahuasi and Muyupampa, (19° 51' S, 63° 43' W), 1430 m, 22 Apr. 2000, *Vásquez & Rivero 3729 a* [LPB!]; ibid.: Villa Vaca Guzman, Serrania del Incahuasi, next to the border to Santa Cruz, 19° 48' 52" S, 63° 43' 10" W, 1540 m, 17 Oct. 2005, *Lliully 341* [HSB!]; ibid.: Villa Vaca Guzman, Serrania del Incahuasi, towards Mmuyupampa, 19° 51' 13" S, 63° 43' 38" W, 1318 m, 19 Oct. 2005, *Lliully 345* [HSB!]; ibid.: Between the summit of Incahuasi and Muyupampa, 19° 40' S, 63° 42' W, 1330 m, 22 Apr. 2000, *Vásquez 3730* [FR!, LPB!, SEL!; FAN!, BGHD!]; ibid.: 13 km W of Cuevo along the road to Iguembe, 20° 24' S, 63° 24' W, 1060 m, 19 July 1982, *Till 102* [WU!; HBV]. **Dpto. Tarija:** Prov. Cercado: Next to the city Tarija, (21° 36' S, 64° 36' W), *Lorenz s.n.* [B!, LPB!]. Prov. O'Connor: Between Lagunita and Entre Ríos, 21° 27' S, 64° 04' W, 1160 m, 14 May 2001, *Vásquez & Ric 4013* [LPB!, SEL!; FAN!]; ibid.: Parque Nacional Tariquia, between Chiquiacá and the valley of San Lucas, 21° 55' S, 64° 09' W, 840 m, 18 May 2001, *Vásquez 4051* [FR!, LPB!, SEL!, VASQ!; BGHD!]; ibid.: 16 km E of Entre Ríos along the road to Palos Blancos, 21° 27' 00" S, 64° 04' 12" W, 1250 m, 15 July 1982, *Till 74* [FR!, WU!; BGB! HBV]. Prov. Gran Chaco: Without precise locality, *Meier s.n.* [B!, FR!; BGB!, BGHD!]; ibid.: Between Villamontes and Palos Blancos, next to Río PilcoMayo, 21° 15' S, 63° 32' W, 320 m, 14 May 2001, *Vásquez & Ric 4003* [LPB!, SEL!, USZ!; FAN!]; ibid.: Between Carapari and Aguayrenda, (21° 54' S, 63° 42' W), 900 m, 20 Oct. 1927, *Troll 446* [B!, M!]; ibid.: 27 km W of Palos Blancos, 21° 27' 00" S, 64° 02' 24" W, 1430 m, 16 July 1982, *Till 79* [WU!]. Prov. Arce: Flor de Oro, next to Bermejo, 22° 44' S, 64° 20' W, 550 m, 21 Dec. 1979, *Coro-Rojas 1442* [LPB!, US!]; ibid.: 111 km from Tarija to Bermejo, along Río Bermejo, between La Merced and Vaden, 22° 12' 23" S, 64° 37' 23" W, 1133 m, 11 Oct. 2006, *Peters 06.0051* [LPB!; BGHD!]; ibid.: 111 km from Tarija to Bermejo, along Río Bermejo, between La Merced and Vaden, 22° 12' 23" S, 64° 37' 23" W, 1133 m, 11 Oct. 2006, *Peters 06.0052* [SEL!]; ibid.: 111 km from Tarija to Bermejo, along Río Bermejo, between La Merced and Vaden, 22° 12' 23" S, 64° 37' 23" W, 1133 m, 11 Oct. 2006, *Peters 06.0053* [FR!; BGHD!]; ibid.: 111 km from Tarija to Bermejo, along Río Bermejo, between La Merced and Vaden, 22° 12' 23" S, 64° 37' 23" W, 1133 m, 11 Oct. 2006, *Peters 06.0054* [BGHD!]; ibid.: 134 km from Tarija to Bermejo, along Río Bermejo, between La Merced and Vaden, 22° 36' 01" S, 64° 25' 32" W, 507 m, 11 Oct. 2006, *Peters 06.0060* [FR!]; ibid.: 200 km from Tarija to Bermejo, along Río Bermejo, between La Merced and Vaden, 22° 41' 46" S, 64° 23' 47" W, 455 m, 11 Oct. 2006, *Peters 06.0061* [LPB!]; ibid.: 200 km from Tarija to Bermejo, along Río Bermejo, between La Merced and Vaden, 22° 41' 46" S, 64° 23' 47" W, 455 m, 11 Oct. 2006, *Peters 06.0062* [SEL!].

ARGENTINA: **Prov. Jujuy:** Dpto. Ledesma: Route to Valle Grande, Tres cruces, 15 Oct. 1964, *Cabrera 15889* [LPB!]; ibid.: Parque Nacional Calilegua, RP 83 to Valle Grande, 4 km before Mesada de las Comenas, 10 Nov. 2000, *Aquino 241* [MCNS!]; ibid.: Sierra de Calilagua, (23° 48' S, 64° 48' W), 800 m, 15 Oct. 1927, *Venturi 5474* [US!]; ibid.: 14 km on RP 83 from RN 34 towards Valle Grande, 23° 43' 38" S, 64° 51' 09" W, 878 m, 29 Nov. 2006, *Peters 06.0139* [BGHD!]; ibid.: 14 km on RP 83 from RN 34 towards Valle Grande, 23° 43' 38" S, 64° 51' 09" W, 878 m, 29 Nov. 2006, *Peters 06.0140* [LIL!; BGHD!]; 14 km on RP 83 from RN 34 towards Valle Grande, 23° 43' 38" S, 64° 51' 9" W, 878 m, 29 Nov. 2006, *Peters 06.0141* [BGHD!]; 14 km on RP 83 from RN 34 towards Valle Grande, 23° 43' 38" S, 64° 51' 09" W, 878 m, 29 Nov. 2006, *Peters 06.0142* [BGHD!]; ibid.: 18 km on RP 83 from RN 34 towards Valle Grande, 23° 42' 14" S, 64° 51' 28" W, 994 m, 29 Nov. 2006, *Peters 06.0146* [LIL!; BGHD!], ibid.: 18 km on RP 83 from RN 34 towards Valle Grande, 23° 42' 14" S, 64° 51' 28" W, 994 m, 29 Nov. 2006, *Peters 06.0147* [BGHD!]. Dpto. San Pedro: RP 56 from San Pedro to San Salvador, 24° 21' 19" S, 65° 01' 38" W, 892 m, 29 Nov. 2006, *Peters 06.0132* [LIL!; BGHD!]; ibid.: RP 56 from San Pedro to San Salvador, 24° 21' 19" S, 65° 01' 38" W, 892 m, 29 Nov. 2006, *Peters 06.0132* [LIL!; BGHD!]; ibid.: RP 56 from San Pedro to San Salvador, 24° 21' 19" S, 65° 01' 38" W, 892 m, 29 Nov. 2006, *Peters 06.0133* [BGHD!]; ibid.: RP 56 from San Pedro to San Salvador, 24° 21' 19" S, 65° 01' 38" W, 892 m, 29 Nov. 2006, *Peters 06.0134* [BGHD!]; ibid.: RP 56 from San Pedro to San Salvador, 24° 21' 19" S, 65° 01' 38" W, 892 m, 29 Nov. 2006, *Peters 06.0135* [BGHD!]; ibid.: RP 56 from San Pedro to San Salvador, 24° 21' 19" S, 65° 01' 38" W, 892 m, 29 Nov. 2006, *Peters 06.0136* [LPB!; BGHD!]; ibid.: RP 56 from San Pedro to San Salvador, 24° 21' 19" S, 65° 01' 38" W, 892 m, 29 Nov. 2006, *Peters 06.0137* [BGHD!]; ibid.: RP 56 from San Pedro to San Salvador, 24° 21' 19" S, 65° 01' 38" W, 892 m, 29 Nov. 2006, *Peters 06.0138* [BGHD!]; ibid.: Sierra de Santa Barbara, (24° 18' S, 64° 30' W), 750 m, 14 Oct. 1929, *Venturi 9721* [GH!, LIL!, MB!, MO!, NY!, US!]; ibid.: Cuesta de las Lajitas, along the road from San Pedro de Jujuy to San Salvador de Jujuy, 24° 06' S, 65° 06' W, 900 m, 7 Feb. 1993, *Till 10083* [B!, FR!, LI!,

LIL!, WU!; BGB!, HBV]. **Dpto. Santa Barbara:** Sierra Santa Barbara, Quebrada de las Termas, 80 km E of San Pedro de Jujuy, (24° 18' S, 64° 30' W), 1500 m, 8 Oct. 1938, *Eyerdam & Beetle 22474* [GH!, MO!]. **Prov. Salta:** Dpto. Orán: Tartagal, Quebradas Río Tartagal, 500 m, 24 Oct. 1924, *Schreiter 3356* [BA!, LIL!]; ibid. Arroyo de Tartagal, 500 m, 27 Sept. 1925, *Schreiter 5035* [F!]; ibid.: Las Tablillas, Vespucio, 700 m, Oct. 1940, *Schreiter 11393* [U!]; ibid.: Angosto del Río Pescado, 650 m, 26 Oct. 1978, *Hadoy A 342* [LIL!]; ibid.: Angosto del Río Pescado, 650 m, 20 Nov. 1967, *Giusti 12133* C [LIL!]; ibid.: Valley of Río Ytau, 50 km W of Manuela Pedraza, 5 km S of San Pedrico, (24° 30' S, 65° 00' W), 1000 m, 29 Oct. 1938, *Eyerdam & Beetle 22751* [K!]; ibid.: Next to Río Blanco, (22° 48' S, 64° 06' W), 600 m, 16 Oct. 1928, *Venturi 7600* [GH!]; ibid.: Tartagal, (22° 30' S, 63° 48' W), 500 m, 29 Sept. 1925, *Schreiter 35* [BA!, K!]; ibid.: Valley of Tartagal, (22° 30' S, 63° 48' W), 500 m, 26 Sept. 1926, *Venturi 5054* [GH!, MA!, US!]. WITHOUT LOCALITY: s.n. [WU 9543!, 9084!, LI 399502!; HBV B181/92 ex BGM 92/2846], s.n. [WU 8555!, 9103!; HBV 294/96, ex living collection Grant Groves], s.r. [WU 10310!; HBV B00B193-1], s.n. [WU 9577!, 9522!, 9017!; HBV B307/96], s.n. [WU 8556!; HBV ex BGP 96/1089], s.n. [WU 9518!, 9519!, 9525!; HBV B98B115-1, -2 & -3 ex BGS], s.n. [WU 9092!, 9584!, 8607!; HBV B97B208-1 ex BGU 96-809], s.n. [WU 8611!, 9596!; HBV B314/96], s.n. [WU 9102!, 8608!, 8609!, 8610!; HBV B97B122-1 -2 & -3 ex BGP], s.n. [B GH-41387!; BGB 289-10-00-80! ex HBV], s.n. [B GH-37114!; BGB 175-01-97-80! ex BGU], s n. [B GH-24630!; BGB 014-77-87-70/73/83/84! ex BGM], s.n. [B GH-R712!; R712!], s.n. [B GH-11251!; BGB s.n.!], s.n. [B GH-43829!; BGB 290-03-00-63/-64! ex HBV], s.n. [M 0124615!; BGM], s.n. [WU 2528!; HBV], s.n. [WU 2529!, 2530!; HBV], s.n. [U 64934!; Cantonspark Baarn 01858], s.n. [FR KS060204-4!; FRP 99-18321-1!], s.n. [FR KS060904-3!; FRP 99-18320-1!], s.r. [FR KS170305-2!; FRP s.n.!], s.n. [LI 399504!, WU 8541!, 8423!, 9060!; HBV B71/96], s.n. [WU 9061!, 9517!; HBV B98B116-1 ex BGS], s.n. [WU 9520!, 9521!; HBV Aussaat 204 ex BGU 96GRD0049], s.n. [WU 9085!; HBV B313/96], s.n. [WU 9059!; HBV B306/96], s.n. [WU 11770!, 11331!; HBV B00B141-1].

Distribution and ecology

Range size: Large.

Countries: BOLIVIA. Dpto. La Paz, Cochabamba, Santa Cruz, Chuquisaca, Tarija.
ARGENTINA: Prov. Jujuy, Salta.

Ecoregions: Tucuman-Bolivian-Forests, Yungas, Inter-Andean-Dry-Forests, Chiquitano-Dry-Forests, Montane Chaco.

Life style & habitats: Dry to humid, deciduous to evergreen lowland and montane forests, especially of the Montane Chaco and some Yungas regions; on rock outcrops within the lowlands (Brazilian shield). Terrestrial or saxicolous, in understory of more or less disturbed forests. In fissures of shaded rocks, on riverbanks and roadsides. On limestone rocks and ledges, granitic outcrops, steep, rocky slopes and loamy hillsides. Humid, deciduous forest with elements of Tucuman-Bolivian-Forests like *Cedrela lilloi* (Meliaceae), *Inga marginata* (Fabaceae), colonies of *Lamprothyrsus hieronimi* (Foaceae) and several species of Solanaceae and Melastomataceae (*Lliully 345* [HSB]). Tucuman-Bolivian-Forest with *Duranta serratifolia* (Verbenaceae), *Baccharis latifolia* (Asteraceae) and *Dodonaea viscosa* (Sapindaceae). Chaco in transitions to Yungas, with *Schinopsis haenkeana* (Anacardiaceae) and *Anadenanthera* sp. (Fabaceae) (*Peters 06.0132* [LIL]). Deciduous forests with frequent cattle-intervention (*Beck & Libermann 9781* [LPB]).

Altitude: 200–2650 m.

Taxonomic delimitation and systematic relationships

Fosterella penduliflora is the most common species of the genus in the field as well as the best represented one in living collections and herbaria. Formerly, the taxa *F. chiquitana* and *F. latifolia* were distinguished from *F. penduliflora*, but morphological delimitation was difficult, because of transitional character states. AFLP data (REX et al. 2007) suggest that plants previously considered as *F. chiquitana* do not represent a monophyletic taxon, but belong to the variable *F. penduliflora*. The same applies to the specimens which previously had been assigned to *F. latifolia*. According to recent molecular analyses, these plants do not represent a monophyletic group either, although they show a certain morphological distinctness (REX et al. 2007). Thus, *F. chiquitana* and *F. latifolia* were considered as synonyms of *F. penduliflora* (PETERS et al. 2008 b).

The specimen *Vásquez & Rivero 3729 a* [LPB!] from the Tucuman-Bolivian-Forests of the Dpto. Chuquisaca, Prov. Luis Calvo shows an abnormal thick indumentum of the abaxial leaf surface. But since all other morphological characters coincide, it has been classified under *F. penduliflora* here.

Vernacular name

In Argentina, *F. penduliflora* is called "Permera" or "Parmera" (*Venturi 5474* [US]).

Fig. 78: *Fosterella penduliflora*. 1. Habit. 2. Inflorescence branch. 3 Peduncle. 4 Leaf. 5 Flower. 6 Petal with stamen attached. 7 Stigmas. Lithograph of *M. Forget s.n.* by G. Atkinson del L. Snelling in STAPF (1924). © Wiley-Blackwell.

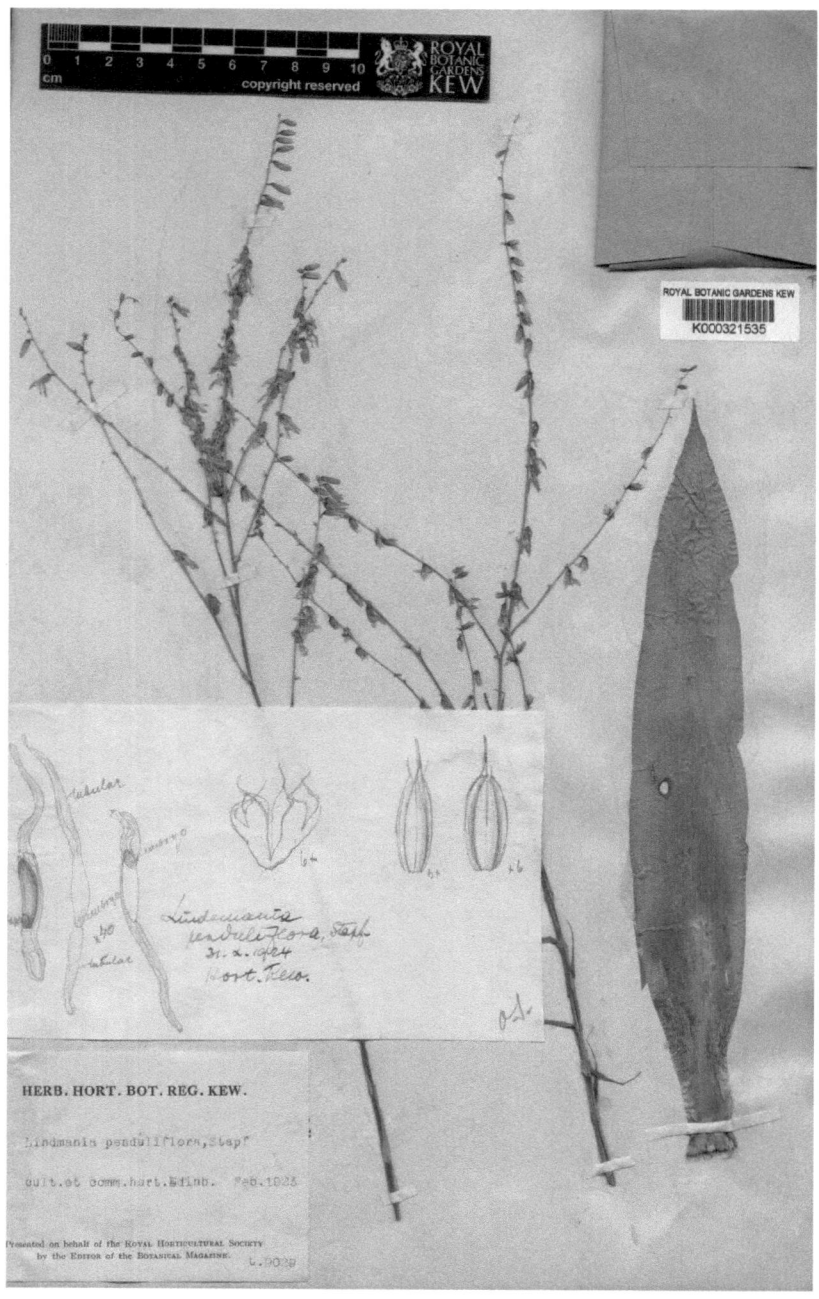

Fig. 79: Lectotype of *Fosterella penduliflora* (C.H. Wright) L.B. Sm. [K]. © The Board of Trustees of the Royal Botanic Gardens, Kew. Reproduced with the consent of the Royal Botanic Gardens, Kew.

Fig. 80: *F. penduliflora (Peters 06.0139)* in the natural habitat.

Fig. 81: Flowers of *F. penduliflora (Peters 06.0039)*.

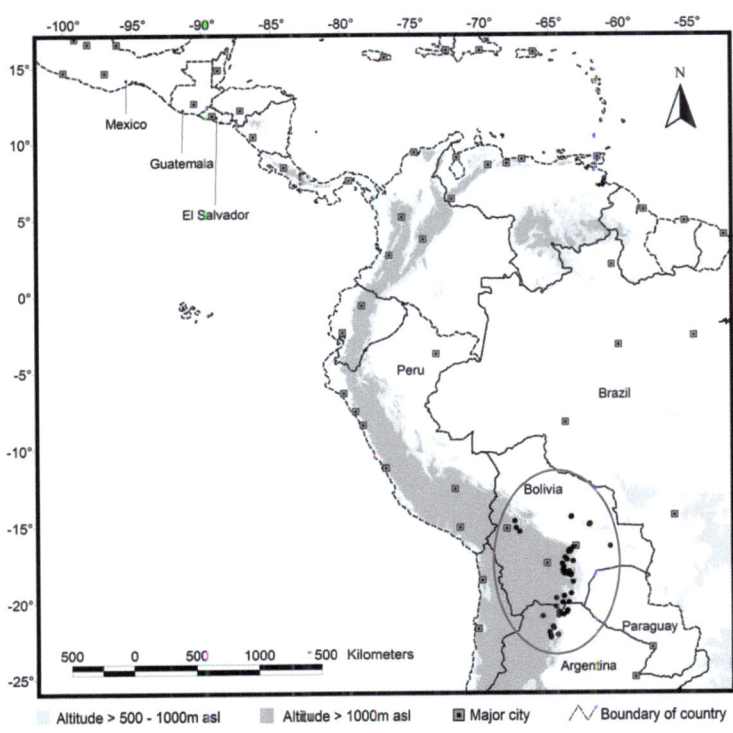

Fig. 82: Distribution of *Fosterella penduliflora*.

113

Fosterella petiolata (Mez) L.B. Sm., Phytologia 7: 172. 1960. Fig. 83–87.

Basionym: *Lindmania petiolata* Mez, Bull. Herb. Boissier II. 4: 864. 1904. TYPE: Perú. Near Tambo Isillame, between Sandia and Chunchusmayo, wood, terrestrial, 1000 m, 23 June 1902, *A. Weberbauer 1210* [Holotype: B!].

Comment on type

As MEZ (1934) already rectified, the type locality is nearby Tambo Isilluma (not "Isillame"), which is located in the Dpto. Puno: Prov. Sandia, approx. at 14° 18' S, 69° 24' W.

Etymology

The epithet refers to the petiolate leaves of this species.

Description

Plant acaulescent, up to 100 cm high. **Leaves** few, up to 10, forming an open, more or less upright rosette. **Sheaths** 20–40 mm wide, entire, whitish, glabrous. **Blades** distinct petiolate, broadly lanceolate, acuminate, 35–75 cm long and to 3.5 cm wide, thin, entire, (petioles 10–35 cm long and 4 mm wide, succulent towards the base, entire), adaxially glabrous, abaxially densely appressed lepidote by peltate trichomes with fimbriate margin. **Peduncle** to 65 cm long, 3–5 mm in diameter, green, arachnoid, glabrescent. **Peduncle bracts** 5–8 cm long, longer than or equalling the internodes, entire, abaxially appressed lepidote, glabrescent. **Inflorescence** paniculate with branches up to 2nd order, 25–40 cm long, 5–10 cm wide, axes green, arachnoid, glabrescent. **Primary bracts** 20–30 mm long, longer than or equalling the sterile base of the branches, entire, abaxially appressed lepidote, glabrescent. **Branches** 8–15, ascending, straight, 3–8 cm long, bearing up to 15 flowers. **Secondary branches** 2–3 cm long, bearing up to 10 flowers. **Floral bracts** 1.5–3 mm long, longer than or equalling the pedicels, entire, abaxially glabrous. **Flowers** spreading, suberect, 3–10 mm apart. **Pedicels** 1–3 mm long. **Sepals** 1–2 mm long, green, glabrous. **Petals** 3–5 mm long, greenish-white, recoiled like watchsprings during anthesis and afterwards. **Filaments** 5 mm long. **Anthers** 2 mm long. **Style** 2 mm long. **Stigmatic complex** simple-erect. **Capsule** ovoid, 3–5 mm long, 3–4 mm wide. **Seeds** filiform, bicaudate, 2 mm long.

Specimens seen

PERÚ: **Dpto. Cuzco:** Prov. La Convención: Railway from Santa Teresa to Aguas Calientes, between hidro electrica and Aguas Calientes, km 119, 13° 09' 55" S, 72° 33' 02" W, 1822 m, 9 Nov. 2006, *Peters 06.0113*. [SEL! (st.), USM! (st.)]; ibid.: Railway from Santa Teresa to Aguas Calientes, between hidro electrica and Aguas Calientes, km 119, 13° 09' 55" S, 72° 33' 02" W, 1822 m, 9 Nov. 2006, *Peters 06.0114* [LPB! (st.), USM! (st.); BGHD!]; ibid.: Railway from Santa Teresa to Aguas Calientes, between hidro electrica and Aguas Calientes, km 119, 13° 09' 55" S, 72° 33' 02" W, 1822 m, 9 Nov. 2006, *Peters 06.0115* [FR! (st.), USM! (st.); BGHD!]; ibid.: Railway from Santa Teresa to Aguas Calientes, between hidro electrica and Aguas Calientes, km 119, 13° 09' 55" S, 72° 33' 02" W, 1822 m, 9 Nov. 2006,

Peters 06.0116 [USM!, (st.), USZ! (st.); BGHD!]; Railway from Santa Teresa to Aguas Calientes, between hidro electrica and Aguas Calientes, km 119, 13° 09' 55" S, 72° 33' 02" W, 1822 m, 9 Nov. 2006, *Peters 06.0117* [LPB! (st.), USM! (st.)]; ibid.: Rosario Mayo, 1050 m, 8 Aug. 1968, *Chavez 323* [CUZ!, US!].

BOLIVIA: **Dpto. La Paz:** <u>Prov. Bautista Saavedra:</u> Pauji-Yuyo, between Apolo and Charazani, 15° 02' S, 68° 29' W, 1200 m, 13 June 1997, *Kessler 10086* [LPB! (st.), SEL! (fr.)]; ibid.: Area Natural de Marejo Integrado Apolobamba, Pauje Yuyo, 15° 02' 12" S, 68° 27' 26" W, 940 m, 7 Sept. 2004, *Cayola 1083* [LPB!, VASQ!]; ibid.: Area Natural de Manejo Integrado Apolobamba; Yurilaya, road to Charazani, 15° 11' 50" S, 68° 36' 00" W, 920 m, 3 Sept. 2004, *Fuentes & Aldana 6682* [LPB!, VASQ!]; ibid.: Area Natural de Manejo Integrado Apolobamba; bridge across Río Charazani, 15° 13' 39" S, 68° 45' 36" W, 1433 m, 3 Sept. 2004, *Fuentes & Aldana 6692* [LPB!, VASQ!]; ibid.: Area Natural de Manejo Integrado Apolobamba, between Carpa and entry to Camata, 15° 12' 59" S; 68° 41' 03" W, 1290 m, 16 Nov. 2003, *Fuentes 6222* [LPB!, USZ!]; ibid.: Pauji-Yuyo, between Apolo and Charazani, 15° 03' S, 68° 29' W, 900 m, 15 June 1997, *Kessler 10190* [LPB!, SEL!] ibid.: Pauji-Yuyo, between Apolo and Charazani, 15° 03' S, 68° 29' W, 900 m, 15 June 1997, *Kessler 10191* [LPB!, SEL!]; ibid: 10 km from Camata to Apolo, 15° 13' S, 68° 41' W, 1300 m, 24 June 1997, *Kessler 10289* [LPB!, SEL!]. <u>Prov. Caranavi:</u> Road from Caranavi to Sapecho, near summit of Serranía Bella Vista, 15° 41' S, 67° 29' W, 1500 m, 5 Aug. 2000, *Krömer & Acebey 1398 a* [USZ!]. <u>Prov. Nor Yungas:</u> 21 km from Coroico to Unduavi, behind Yolosa, 16° 14' 12" S, 67° 47' 26" W, 1723 m, 21 Oct. 2006, *Peters 06.0095* [SEL!]; ibid.: 21 km from Coroico to Unduavi, behind Yolosa, 16° 14' 12" S, 67° 47' 26" W, 1723 m, 21 Oct. 2006, *Peters 06.0096* [LPB!]; ibid.: 21 km from Coroico to Unduavi, behind Yolosa, 16° 14' 12" S, 67° 47' 26" W, 1723 m, 21 Oct. 2006, *Peters 06.0097* [FR!; BGHD!]; ibid.: 21 km from Coroico to Unduavi, behind Yolosa, 16° 14' 12" S, 67° 47' 26" W, 1723 m, 21 Oct. 2006, *Peters 06.0098* [SEL!]; ibid.: 21 km from Coroico to Unduavi, behind Yolosa, 16° 14' 12" S, 67° 47' 26" W, 1723 m, 21 Oct. 2006, *Peters 06.0099* [LPB!]; ibid.: Cotapata National Park, trails around Estación Biológica de Tunquini, 16° 12' S, 67° 51' W, 1550 m, 25 Aug. 1999, *Krömer & Acebey 763* [FR!, SEL!]; ibid.: Cotapata National Park, trails around Estación Biológica de Tunquini, 16° 11' S, 67° 53' W, 1650 m, 22 Sept. 2000, *Krömer & Acebey 1594* [SEL!, USZ!]; ibid.: 15 km on the road from Chuspipat to Yolosa, 16° 17' S, 67° 48' W, 1800 m, 26 Sept. 1999, *Krömer & Acebey 891* [FR!, SEL!]; ibid.: 27 km on the road from Chuspipata to Yolosa, 16° 17' S, 67° 48' W, 1250 m, 26 Sept. 1999, *Krömer & Acebey 893* [FR!, SEL!]; ibid.: Cotapata National Park, 16° 12' S, 67° 50' W, 1300 m, 5 Oct. 1999, *Krömer & Acebey 902* [FR!, SEL!]; ibid: Cotapata National Park, Estacion Biologica de Tunquini, 16° 11' S, 67° 52' W, 1680 m, 27 Sept. 2001, *Maldonado BMA 201* [LPB!].

WITHOUT LOCALITY: *Leme 2414* [WU 9464!; HBV B97B369-1], *s.n.* [WU 9074!, 10237!, 9523!; HBV B97B183-1 ex living collection Vandervart s.n.], *s.n.* [SEL 090011!, 090012!, 090013!, 090019!; MSBG 1995-0007 ex living collection Baensch s.n.].

Distribution and ecology

Range size: Small.

Countries: PERÚ. Dpto. Cuzco. BOLIVIA. Dpto. La Paz.

Ecoregions: Peruvian and Bolivian Yungas.

Life style & habitats: Humid, evergreen montane forests; understory of more or less disturbed forests. Terrestrial or saxicolous, along roadsides, on steep slopes, quite abundant. Semideciduous forest on dry slopes with *Anadenanthera colubrina* (Fabaceae) (*Fuentes 6222* [LPB]). Humid, evergreen montane forests with *Oenocarpus bataua* (Arecaceae), *Matayba* sp. (Sapindaceae) and *Cyathea* sp. (Cyatheaceae), Melastomataceae, Lauraceae and Myrtaceae (*Cayola 1083* [LPB]). Evergreen forest, slightly disturbed, with *Attalea* sp. and *Iriartea* sp. (Arecaceae) (*Kessler 10190* [LPB]).

Altitude: 900–1800 m.

Taxonomic delimitation and systematic relationships

Fosterella petiolata is similar to *F. kroemeri*, but differs in the clearly petiolate leaves, the slightly arachnoid inflorescence, longer peduncle bracts and primary bracts. *Fosterella petiolata* seems to be morphologically quite variable, but ecologically well defined, being fairly abundant at a certain altitudinal belt in the La Paz Yungas. In Cotapata National Park, large populations have been observed.

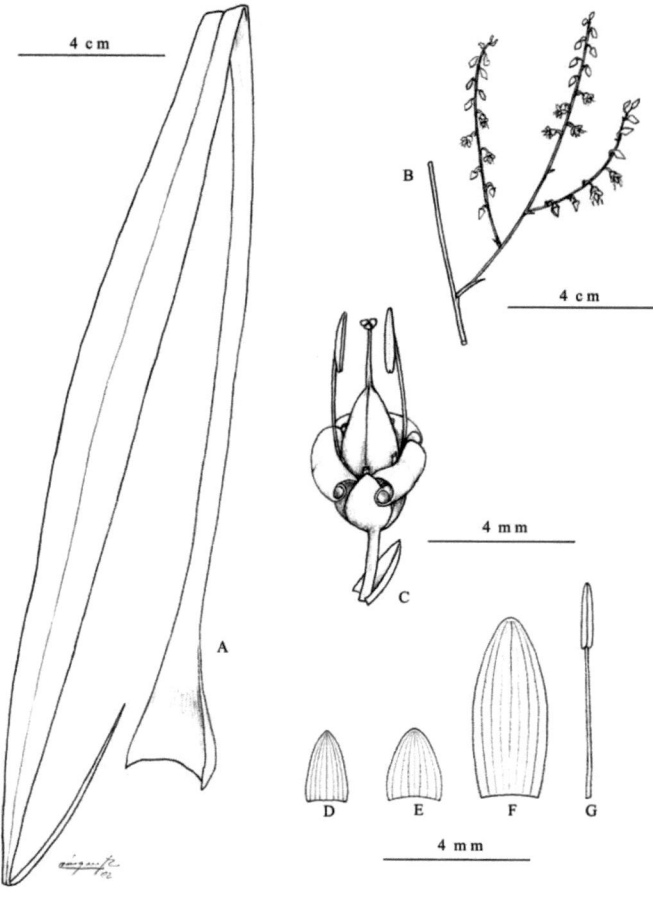

Fig. 83: *Fosterella petiolata*. A Leaf. B Inflorescence branch and primary bract. C Flower. D Floral bract. E Sepal. F Petal. G Stamen. Drawing of *Krömer & Acebey 1594* by R. Vásquez in IBISCH et al (2002). © Selbyana.

Fig. 84: Holotype of *Fosterella petiolata* (Mez) L.B. Sm. [B].

Fig. 85: *F. petiolata (Peters 06.0095)* in the natural habitat.

Fig. 86: Flowers of *F. petiolata (Peters 06.0095)*.

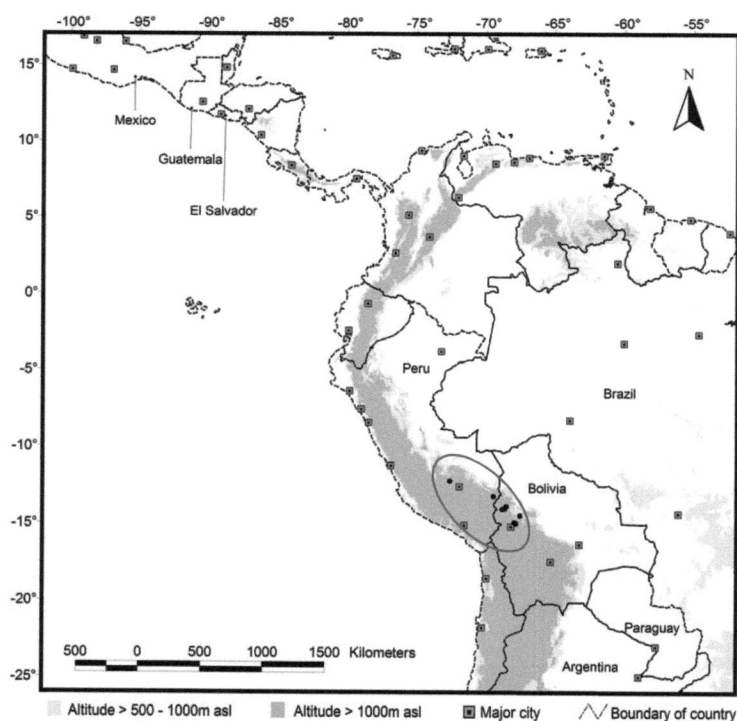

Fig. 87: Distribution of *Fosterella petiolata*.

Fosterella rexiae Ibisch, R. Vásquez & E. Gross, Selbyana 23 (2): 213. 2002. Fig. 88–92.

TYPE: Bolivia. Dpto. La Paz: Prov. Caranavi, 25 km from Caranavi to Yolosa, 15° 57' S, 67° 34' W, 830 m, 15 Feb. 2000 (cultivated specimen flowered in September 2001), R. *Vásquez & G. Gerlach 3673* [Holotype: LPB!]. Bolivia. Dpto. La Paz: Prov. Caranavi, road from Caranavi to Yucumo, between the summit and Sapecho, 13° 10' S, 67° 29' W, 1310 m, 14 Feb. 2000, R. *Vásquez & G. Gerlach 3666* [Paratype: USZ!].

Comment on type

A Clonotype is cultivated in the living collection of the Fundación Amigos de la Naturaleza, Santa Cruz (Bolivia).

Etymology

This species is dedicated to the German biologist Martina Rex (*1976), who prepared a molecular phylogeny of the genus *Fosterella* L.B. Sm.

Description

Plant caulescent, up to 90 cm high. **Leaves** few, 10–15, forming an open, arched rosette. **Sheaths** to 25 mm wide, entire, whitish, glabrous. **Blades** linear to narrowly lanceolate, acuminate, narrowed towards the base, 20–30 cm long and 1.7 cm wide, succulent and serrate towards the base, adaxially lepidote towards the base, abaxially densely appressed lepidote by peltate trichomes with fimbriate margin. **Peduncle** to 55 cm long, 3 mm in diameter, green or reddish, slightly arachnoid. **Peduncle bracts** 5 cm long, longer than or equalling the internodes, entire, abaxially slightly arachnoid, glabrescent. **Inflorescence** paniculate with branches up to 2nd order, to 50 cm long and 15 cm wide, axes green to reddish, slightly arachnoid, glabrescent. **Primary bracts** 15 mm long, longer than the sterile base of the branches, entire, abaxially slightly arachnoid, glabrescent. **Branches** 15–18, inclined, arcuate, 8–10 cm long, bearing up to 20 flowers. **Secondary branches** 3–5 cm long, bearing 6–10 flowers. **Floral bracts** 2 mm long, longer than the pedicels, abaxially slightly arachnoid, glabrescent. **Flowers** spreading, erect, sessile, 3–5 mm apart. **Sepals** 3 mm long, green to reddish, sparsely villous, glabrescent. **Petals** 5 mm long, white, recoiled like watchsprings during anthesis and afterwards. **Filaments** 5 mm long. **Anthers** 1.5 mm long. **Style** 5 mm long. **Stigmatic complex** simple-erect. **Capsule** ovoid, 4 mm long, 2–3 mm wide. **Seeds** filiform, bicaudate, 2 mm long.

Specimens seen

Hitherto this new species is recorded from the type localites only.

Distribution and ecology

Range size: Very small.
Country: BOLIVIA. Dpto. La Paz.
Ecoregions: Sub-Andean-Amazon-Forests, Yungas.
Life style & habitats: Humid, evergreen lowland and montane rain forests.
Altitude: 800–1300 m.

Taxonomic delimitation and systematic relationships

Fosterella rexiae is close to *F. albicans* but differs in the flatter rosette and less pronounced indumentum of the abaxial leaf surfaces, a laxer, less densely flowered and less arachnoid inflorescence with longer and more slender branches and shorter floral bracts, sepals and petals. Fom *F. pearcei* it differs in wider leaf blades and floral bracts which are longer than the sterile base of the branches.

Molecular data have shown that *F. rexiae* is closely related to, but distinct from *F. albicans* (REX et al. 2007), both belonging to the morphologically heterogeneous *albicans*-group.

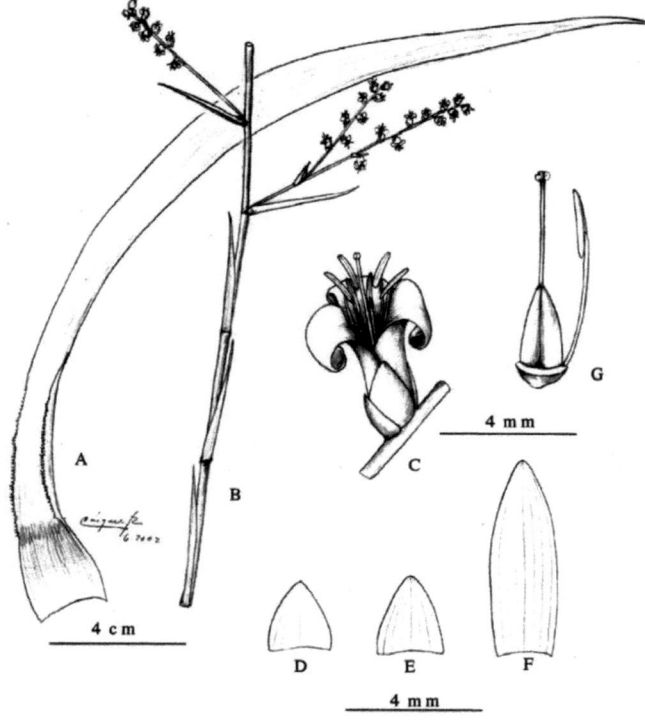

Fig. 88: *Fosterella rexiae*. A Leaf. B Scape, scape bracts, primary bracts and branches. C Flower. D Floral bract. E Sepal. F Petal. G Gynoecium and stamen. Drawing of *Vásquez & Gerlach 3673* by R. Vásquez in IBISCH et al. (2002). © Selbyana.

Fig. 89: Holotype of *Fosterella rexiae* Ibisch, R. Vásquez & E. Gross [LPB].

Fig. 90: *F. rexiae (Vásquez & Gerlach 3673)*.
Photo: P. Ibisch.

Fig. 91: Flowers of *F. rexiae (Vásquez & Gerlach 3673)*. Photo: P. Ibisch.

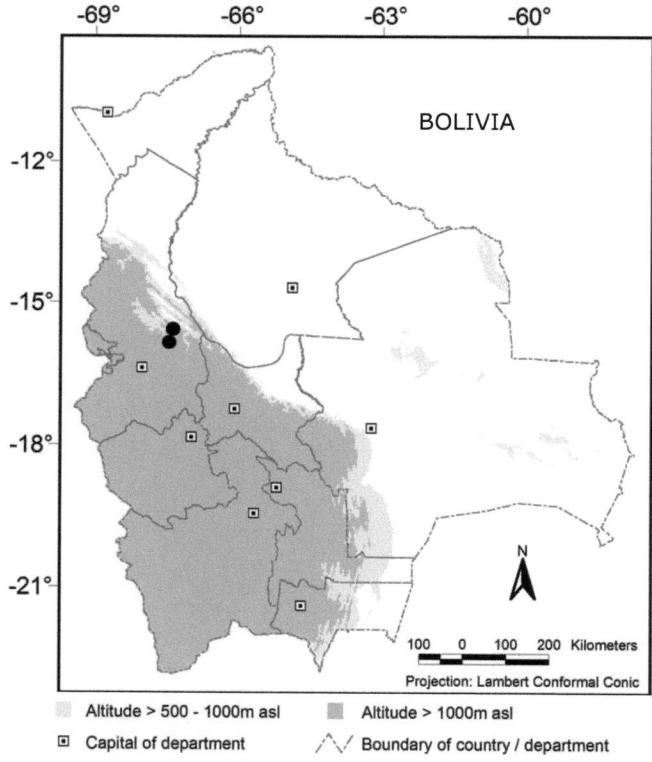

Fig. 92: Distribution of *Fosterella rexiae*.

Fosterella robertreadii Ibisch & J. Peters, Selbyana 29 (2): 192. 2008. Fig. 93–96.

> TYPE: Perú. Dpto. Cusco: Prov. Quillabamba, Road from Quillabamba to Echarate, along the Rio Vilcanota, next to Maranura, 13° 00' 46" S, 72° 38' 32" W, 1168 m, 12 Nov. 2006, *J. Peters 06.0131* [Holotype: SEL!; Isotype: USM!]. Perú. Dpto. Cuzco: Prov. Quillabamba, Road from Quillabamba to Echarate, along the Rio Vilcanota, next to Sietetinajas, 12° 46' 08" S, 72° 38' 02" W, 968 m, 12 Nov. 2006, *J. Peters 06.0126* [Paratype: LPB!, USM!].

Comment on type

A Clonotype is cultivated in the Botanical Gardens Heidelberg (Germany).

Etymology

This species is dedicated to the American Bromeliad and *Fosterella* specialist Robert William Read (1931–2003), who was the first who identified this species as new, even though he never published his discovery.

Description

Plant acaulescent, up to 100 cm high. **Leaves** many, up to 30, forming a dense, arched rosette. **Sheaths** to 50 mm wide, entire, whitish, glabrous. **Blades** narrowly linear, slightly involute, to 80 cm long and 2 cm wide, succulent towards the base, entire, (rarely some minute spines at the base), adaxially slightly appressed lepidote, abaxially densely appressed lepidote by peltate trichomes with fimbriate margin. **Peduncle** to 70 cm long, 5 mm in diameter, green, arachnoid, glabrescent. **Peduncle bracts** to 6 cm long, considerably longer than internodes, entire, sometimes reddish, abaxially densely appressed lepidote. **Inflorescence** paniculate, with lateral branches up to 3rd order, to 40 cm long and 15 cm wide, axis green, arachnoid, glabrescent. **Primary bracts** 20 mm long, longer than the sterile base of the branches, entire, abaxially appressed lepidote. **Branches** up to 30, ascending, straight, to 20 cm long, bearing up to 30 flowers. **Secondary branches** to 10 cm long, bearing up to 15 flowers. **Floral bracts** 1 mm long, shorter than pedicels, entire, sometimes reddish, abaxially glabrous. **Flowers** spreading, pendulous, 3–5 mm apart. **Pedicels** to 2 mm long. **Sepals** 1.5 mm long, green, sometimes reddish, glabrous. **Petals** 4 mm long, white, recoiled like watchsprings during anthesis and afterwards. **Filaments** 2 mm long. **Anthers** 1 mm long. **Style** 2 mm long. **Stigmatic complex** simple-erect. **Capsule** ovoid, 3 mm long, 2 mm wide. **Seeds** filiform, bicaudate, 2 mm long.

Specimens seen

PERÚ: Without precise locality: Sept. 1925, *Dieh. 2437* [F!]; Sept. 1925, *Diehl 2513* [F!]. **Dpto. Cuzco**: Valley of Río Vilcanota, 1200 m, Sept. 1967, *Rauh 20866* [B!, CUZ!, HEID!, LPB!, WU!; BGHD!, BGOS! ex BGM ex living collection Marnier-Lapostolle ex BGHD!, HBV ex HEID!, BGB! ex BGHD!]; ibid.: Pumachaca, Santa Ana Valley, 1400 m, Oct.

1931, *Herrera 3316* [GH!]. Prov. La Convención: Quillabamba, Cocalpampa, Chaullay, Maranura, Quintalpata, 1300 m, 29 Dec. 1986, *Nuñez 6787 a* [CUZ! (fr.), USM! (fr.)]; ibid.: Road from Santa Maria to Quillabamba, along Río Vilcanota, 12° 56' 20" S, 72° 39' 34" W, 1155 m, 11 Nov. 2006, *Peters 06.0118* [USM! (st.); BGHD!]; ibid.: Road from Santa Maria to Quillabamba, along Río Vilcanota, 12° 56' 20" S, 72° 39' 34" W, 1155 m, 11 Nov. 2006, *Peters 06.0119* [SEL!; BGHD!]; ibid.: Road from Santa Maria to Quillabamba, along Río Vilcanota, 12° 56' 20" S, 72° 39' 34" W, 1155 m, 11 Nov. 2006, *Peters 06.0120* [FR! (st.), USM!; BGHD!]; ibid.: Road from Santa Maria to Quillabamba, along Río Vilcanota, 12° 56' 20" S, 72° 39' 34" W, 1155 m, 11 Nov. 2006, *Peters 06.0121* [FR!, USM! (st.); BGHD!]; ibid.: Road from Santa Maria to Quillabamba, along Río Vilcanota, 12° 57' 19" S, 72° 40' 19" W, 1172 m, 11 Nov. 2006, *Peters 06.0122* [USM! (st.)]; ibid.: Road from Quillabamba to Echarate, along Río Vilcanota, next to Echarate, 12° 46' 06" S, 72° 35' 14" W, 902 m, 12 Nov. 2006, *Peters 06.0124* [LPB! (st.), USM!; BGHD!]; ibid.: Road from Quillabamba to Echarate, along Río Vilcanota, next to Echarate, 12° 46' 06" S, 72° 35' 14" W, 902 m, 12 Nov. 2006, *Peters 06.0125* [LPB!, USM! (st.)]; ibid.: Road from Quillabamba to Echarate, along Río Vilcanota, next to Maranura, 13° 00' 46" S, 72° 38' 32" W, 1168 m, 12 Nov. 2006, *Peters 06.0129* [FR!, USM! (st.); BGHD!]; ibid.: Road from Quillabamba to Echarate, along Río Vilcanota, next to Maranura, 13° 00' 46" S, 72° 38' 32" W, 1168 m, 12 Nov. 2006, *Peters 06.0130* [SEL! (st.), USM!; BGHD!].
BOLIVIA: **Dpto. La Paz:** Solacama, between Chulumani and Irupana, (16° 24' S, 67° 28' W), 1100 m, *Mijagawa s.n.* [FR!, HEID!, LPB!]. Prov. Murillo: 44 km below the lake Zongo, vicinity of Cahua, 16° 03' 00" S, 68° 01' 12" W, 1200 m, 12–15 Sept. 1983, *Solomon 10836* [LPB!, MO!, NY!, SEL!]. Prov. Caranavi: 44 km from Caranavi to Sapecho, Serranía Bella Vista, 15° 40' S, 67° 29' W, 1300 m, 29 Aug. 1997, *Kessler 11556* [LPB! (fr.), SEL! (fr.)]. Prov. Nor Yungas: Polo-Polo, near Coroico, (16° 01' 48" S, 67° 42' 00" W), 1100 m, Oct. 1912, *Buchtien 3674* [GH!, NY!, US!]. WITHOUT LOCALITY: *s.n.* [B 16540!; BGB 115-19-83-80! ex BGHH 81-G-496! ex living collection Hemker], *s.n.* [SEL 090212!, 090010!, 026610!; MSBG 1978-0905 ex living collection Cathcart], *s.n.* [SEL 063793!; MSBG s.n.], *s.n.* [HEID 600234!, 602060!, 602061!].

Distribution and ecology

Range size: Very small.
Countries: PERÚ. Dpto. Cuzco. BOLIVIA. Dpto. La Paz.
Ecoregions: Peruvian and Bolivian Yungas.
Life style & habitats: Slightly disturbed, humid, evergreen lowland and montane forests, along roadsides.
Altitude: 900–1400 m.

Taxonomic delimitation and systematic relationships

Fosterella robertreadii is similar to *F. rojasii* but has longer, linear leaf blades, which are adaxially slightly lepidote towards the base, the arachnoid peduncle, longer primary bracts and shorter pedicels. From *F. graminea* it differs in the almost entire leaf blades which are adaxially slightly lepidote towards the base and the arachnoid peduncle and not secund flowers.

Robert W. Read had already identified this species as new by labelling the living plant Acc. No. 1978-0905 from Marie Selby Botanical Gardens as "sp. nov.", but never published it. By revising this specimen (herbarium material), living plants in the Botanical Gardens Heidelberg collected by Werner Rauh (*Rauh 20866*) and studying specimens in its natural habitat, very close to the locality, where Rauh already collected this species 40 years ago, this taxon had been 're-discovered' (PETERS et al. 2008 b).

According to molecular data, several *F. robertreadii*-specimens form a well supported subclade within the "*albicans*"-group, which furthermore comprises *F. rexiae, F. caulescens, F. albicans, F. kroemeri* and *F. graminea.*

Fig. 93: Holotype of *Fosterella robertreadii* Ibisch & J. Peters [SEL].

Fig. 94: *F. robertreadii (Peters 06.0124)* in the natural habitat.

Fig. 95: Flowers of *F. robertreadii (Peters 06.0118)*.

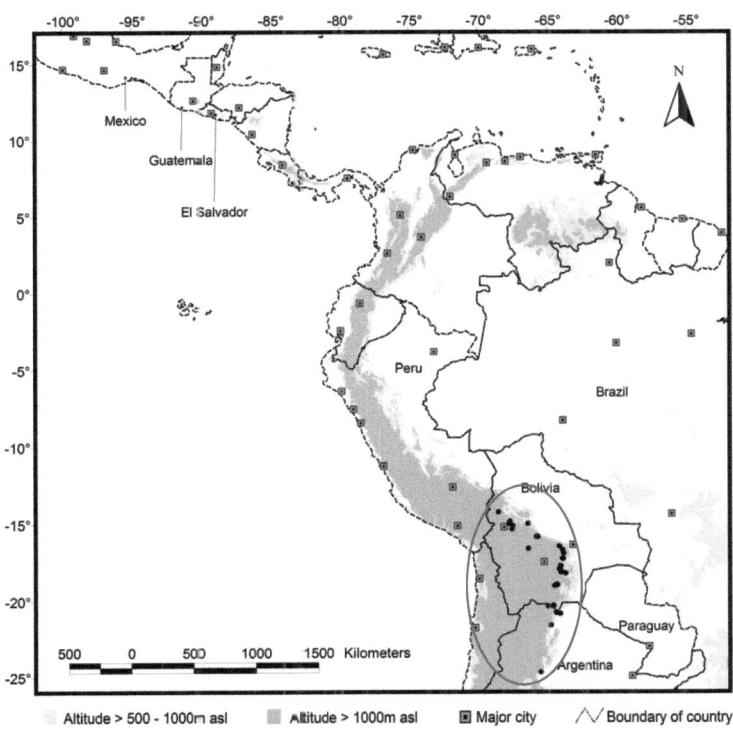

Fig. 96: Distribution of *Fosterella robertreadii*.

Fosterella rojasii (L.B. Sm.) L.B. Sm., Phytologia 7: 172. 1960. Fig. 97–99.

Basionym: *Lindmania rojasii* L.B. Sm., Revista Argent. Agron. 7: 162. 1940. TYPE: Paraguay. Dpto. Amambay: Cerro Corá, Sierra de Amambay, between shady rocks, Mar. 1934, *T. Rojas 6771* [Holotype: GH!; Isotype: BA!].

Comment on type

The Sierra de Amambay is located approx. at 22° 20' S, 55° 48' W.

Etymology

This species is dedicated to the Paraguayan botanist Teodoro Rojas (1877–1954), collector of the type specimen.

Description

Plant acaulescent, up to 50 cm high. **Leaves** few, 8–15, forming an open, flat rosette. **Sheaths** to 30 mm wide, entire, whitish, glabrous. **Blades** linear to narrowly lanceolate, long-acuminate, 20–30 cm long and 1.5–2.5 cm wide, thin, entire, adaxially glabrous, abaxially covered by a thick layer of interwoven, peltate trichomes with fimbriate margin. **Peduncle** 20–25 cm long, to 3 mm in diameter, green, glabrous. **Peduncle bracts** 4–7 cm long, longer than or equalling the internodes, entire, abaxially covered by a layer of interwoven trichomes. **Inflorescence** paniculate with branches up to 2nd order, 25 cm long and 15 cm wide, axes green, glabrous. **Primary bracts** 5 mm long, much shorter than the sterile base of the branches, entire, abaxially interwoven lepidote, glabrescent. **Branches** 6–8, ascending, straight, 10–15 cm long, bearing up to 30 flowers. **Secondary branches** to 5–15 cm long, bearing up to 25 flowers. **Floral bracts** 1–2 mm long, much shorter than the pedicles, entire, abaxially glabrous. **Flowers** secund, pendulous, 5–7 mm apart. **Pedicels** to 5 mm long. **Sepals** 2 mm long, green, glabrous. **Petals** 5–6 mm long, white, recoiled like watchsprings during anthesis and afterwards. **Filaments** 4 mm long. **Anthers** 1.5 mm long. **Style** 3 mm long. **Stigmatic complex** simple-erect. **Capsule** ovoid, 4–5 mm long and 3 mm wide. **Seeds** filiform, bicaudate, 2 mm long.

Specimens seen

PARAGUAY: **Dpto. Amambay:** Estancia la Serrana, buffer zone of Parque Nacional Cerro Cora, Cerro Lorito, 22° 41' 12" S, 55° 59' 17" W, 14 June 1996, *Zardini & Cardozo 45103* [MO!, SEL!]; ibid.: Estancia Don Juancito: Cerro Ysau, 22° 40' 12" S, 59° 09' 35" W, 270 m, 12 June 1996, *Zardini & Cardozo 44974* [NY! (st.), SEL!]; ibid.: Parque Nacional Cerro Corá, Cerro Muralla, 22° 30' S, 56° 00' W, 24 Oct. 1994, *Krapovickas 45675* [K!; KEW 1981-406]; ibid.: Parque Nacional Cerro Cora, Nemore Collino, Cerro Lorito, 440 m, 8 Dec. 1978, *Bernardi 18987* [NY!]; ibid.: Parque Nacional Cerro Corá, summit of Cerro Muralla, Oct. 1980, *Casas & Molero* [NY!]; ibid.: Parque Nacional Cerro Corá, vicinity of Cerro Muralla 22° 39' S, 56° 03' W, 300 m, 7 Feb. 1982, *Solomon 6781* [MO!, NY!]; ibid.: Parque Nacional Cerro Cora, 22° 40' S, 56° 05' W, 300 m, 18 Mar. 1983, *Hahn 1211* [MO!]; ibid.: Parque Nacional Cerro Cora, 22° 40' S, 56° 05' W, 300 m, 7 May 1984, *Hahn 2384* [MO!]; ibid.: Parque Nacional Cerro Cora, Cerro Murrallo, 6 Jan. 1988, *Soria & Zardini 1972* [MO!].

Distribution and ecology

Range size: Very small.
Country: PARAGUAY. Dpto. Amambay.
Ecoregion: Paraná-Paraíba-Interior-Forests.
Life style & habitats: Pre-Cambrian rock outcrops in more or less disturbed, semideciduous forests. Saxicolous on rocky slopes, among shaded rocks.
Altitude: 250–450 m.

Taxonomic delimitation and systematic relationships

Fosterella rojasii is similar to *F. penduliflora*, but differs in the adaxial leaf surface, which bears a dense layer of interwoven trichomes, shorter floral bracts and petals and a compact stigmatic complex. *Fosterella rojasii* is known from Paraguayan Chaco only, and up to now, no molecular data is available.

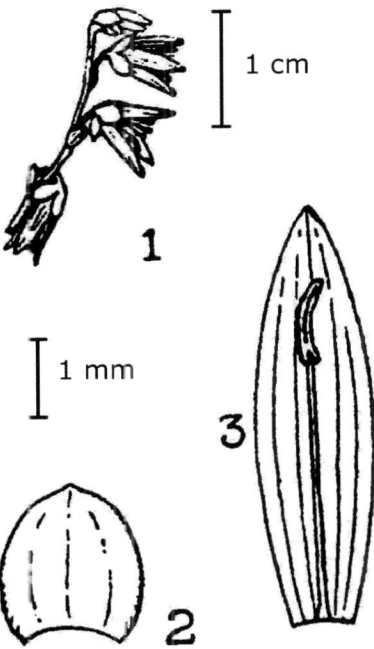

Fig. 97: *Fosterella rojasii*. 1 Inflorescence branch. 2 Sepal. 3 Petal and stamen. Drawing of *Rojas 6771* in SMITH (1940). © Sociedad Argentina de Agronomía.

Fig. 98: Holotype of *Fosterella rojasii* (L.B. Sm.) L.B. Sm. [GH].

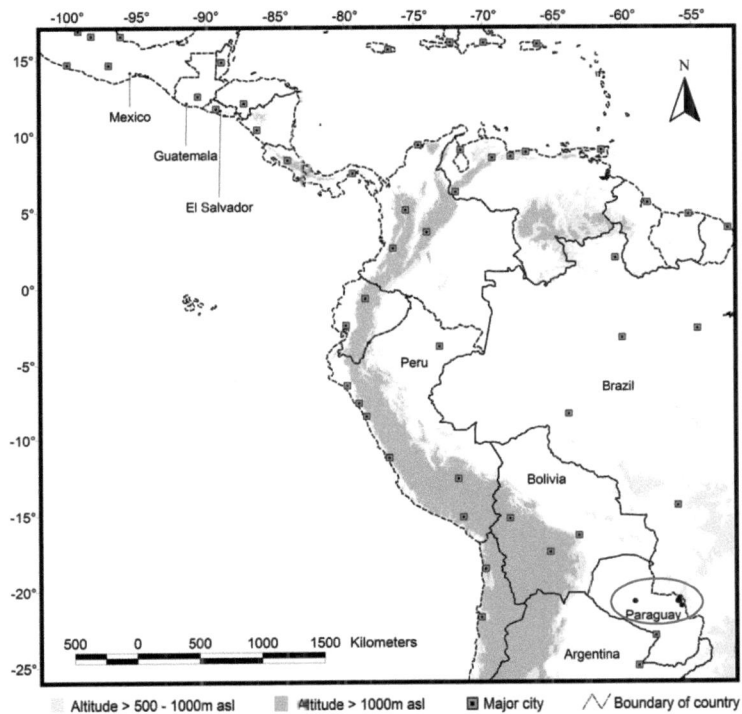

Fig. 99: Distribution of *Fosterella rojasii*.

Fosterella rusbyi (Mez) L.B. Sm., Phytologia 7: 172. 1960. Fig. 100–104.

Basionym:	*Lindmania rusbyi* Mez, Bot. Jahrb. Syst. 30, Beibl. 67: 6. 1901. TYPE: Bolivia. Dpto. La Paz: Yungas, *M. Bang 2571* [Lectotype: B!; Isotypes: F!, GH!, K!, MO!, NY!, US!).
=	*Fosterella elata* H. Luther, Selbyana 5: 310. 1981. TYPE: Bolivia. Dpto. La Paz: Prov. Sud Yungas, terrestrial on rocky hillsides along the Río Unduavi, 1500 m, 2 Feb. 1980, *C. Luer, J. Luer, R. Vásquez & R. Lara s.n.* [Holotype: SEL!].

Comment on type

The type locality of *Fosterella elata* H. Luther is located at approx. 16° 20' S, 67° 49' W. A Clonotype of *F. elata* H. Luther is cultivated in Marie Selby Botanical Gardens, Sarasota (Florida).

Etymology

This species is dedicated to Henry Hurd Rusby (1855–1940), Professor of Botany and Materia Medica at the College of Pharmacy at Columbia University and Director of the Mulford Biological Exploration of the Amazon Basin.

Description

Plant acaulescent, up to 200 cm high. **Leaves** few, 10–15, forming an open, flat rosette. **Sheaths** 50–60 mm wide, entire, whitish, glabrous. **Blades** linear to narrowly lanceolate, acuminate, narrowed towards the base, 20–45 cm long and 1–6 cm wide, succulent and serrate towards the base, undulate, adaxially lepidote towards the base, abaxially covered by a thick layer of interwoven, peltate trichomes with fimbriate margin. **Peduncle** 30–120 cm long, 3–5 mm in diameter, green to reddish, glabrous. **Peduncle bracts** 4–10 cm long, longer than the internodes, entire, abaxially interwoven lepidote. **Inflorescence** paniculate with branches up to 2nd order, 20–50 cm long and 8–25 cm wide, axes green, glabrous. **Primary bracts** 6–10 (25) mm long, shorter than or equalling the sterile base of the branches, entire, abaxially lepidote, glabrescent. **Branches** 10–15, inclined, arcuate, 8–15 cm long, bearing up to 15 flowers. **Secondary branches** 5–8 cm long, bearing up to 10 flowers. **Floral bracts** 2–3 mm long, shorter than or equalling the pedicels, entire, abaxially glabrous. **Flowers** secund, pendulous, 5–10 mm apart. **Pedicels** 1.5–2 mm long. **Sepals** 2–3 mm long, green, glabrous. **Petals** 7 mm long, frequently greenish, cream or rose, recoiled like watchsprings during anthesis and afterwards. **Filaments** 3 mm long. **Anthers** 1 mm long. **Style** 2 mm long. **Stigmatic complex** simple-erect. **Capsule** ovoid, 5 mm long, 3 mm wide. **Seeds** filiform, bicaudate, 2 mm long.

Specimens seen

BOLIVIA: Without precise locality, *Vogel 880874* [B!; BGB 203-09-98-60!]. **Dpto. La Paz:** Without precise locality, *Reichle s.n.* [LPB!, SEL!]; ibid.: Without precise locality, Feb. 2000, *Nowicki 2066* [LPB!, SEL!]. Prov. Franz Tamayo: Parque Nacional Madidi, Wilinculín, upstream Río Mojos, between Mojos and Charopampa, 14° 33' 28" S, 68° 50' 20"

W, 1350 m, 4 July 2005, *Fuentes 9396* [LPB!]; ibid. Parque Nacional Madidi, Río Tuichi, Virgen del Rosario, 14° 35' 40" S, 68° 41' 20" W, 900 m, 25 Nov. 2005, *Araujo 2278* [LPB!]; ibid.: Parque Nacional and Area Natural de Manejo Integrado Madidi, N of Apolo, 14° 19' 31" S, 68° 33 59" W, 767 m, 25 May 2005, *Uzquiano 185* [LPB!]; ibid.: Parque Nacional and Area Natural de Manejo Integrado Madidi, N of Apolo, 14° 18' 50" S, 68° 32' 28" N, 721 m, 8 June 2005, *Uzquiano 207* [LPB!]; ibid.: Area Natural de Manejo Integrado Madidi, 3 km N of Unapa, 14° 32' 35" S, 68° 29' 38" W, 969 m, 3 Sept. 2004, *Cayola 874* [LPB]; ibid.: Parque Nacional Madidi, between Apolo and Azariamas, towards Arroyo Pintata, 14° 27' 55" S, 68° 32' 35" W, 1015 m, 24 Feb. 2003, *Paniagua 5610* [LPB! (st.)]; ibid.: Parque Nacional Madidi, between Virgen del RosaRío and Pata, 14° 35' 48" S, 68° 40' 59" W, 1150 m, 10 Nov. 2003, *Fuentes 5888* [LPB!, MO!]; ibid.: Parque Nacional Madidi, Río Querdeque, 16° 00' 28" S, 67° 46' 25" W, 300 m, 7 Feb. 2002, *Fuentes 3758* [LPB! (st.),SEL! (st.)]; ibid.: 10 km NW of Apolo, Río Bilipisa, 14° 36' S, 68° 27' W, 1100 m, 4 July 1997, *Kessler 11008* [LPB!, SEL!]; ibid.: Hills between Río Machariapo and Río Ubito, 14° 25' S, 68° 31' W, 900 m, 22 July 1993, *Kessler 4100* [LPB!, SEL!]. Prov. Bautista Saavedra: 12 km from Camata to Apolo, 15° 13' S, 68° 41' W, 1250 m, 26 June 1997, *Kessler 10363* [LPB!]. Prov. Muñecas: Consata, (15° 18' S, 68° 24' W), 1200 m, *Besse s.n.* [SEL!; MSBG]. Prov. Larecaja: Dry valley next to Consata, 15° 24' 36" S, 68° 32' 24" W, 1800 m, 8 Aug. 2002, *Nowicki & Deichmann 2356* [FR!; FAN!]. Prov. Caranavi: Between Caranavi and Coroico, 16° 01' 48" S, 67° 40' 00" W, 1100 m, 31 Oct. 1997, *Krömer & Acebey 124* [SEL!]; ibid.: Between Caranavi and Palos Blancos, 1 km behind the bridge across Río Carrasco, towards Alto Lima, 15° 48' S, 67° 28' W, 900 m, 27 Sept. 2001, *Jiminez & Bach 652* [LPB!, SEL!]; ibid.: Between Caranavi and Palos Blancos, 5 km behind San Lorenzo towards Alto Lima, 15° 48' S, 67° 28' W, 1280 m, 27 Sept. 2001, *Jiminez & Bach 659* [LPB!, SEL!]; ibid.: Between Sapecho and Quiquibey, 15° 27' S, 67° 10' W, 900 m, 27 Oct. 2002, *Vásquez 4667 a* [LPB!]; ibid.: 16 km from Caranavi to Sapecho, 15° 46' S, 67° 30' W, 950 m, 26 Oct. 2002, *Vásquez 4655* [FR!, VASQ!; FAN!]; ibid.: 2 km from Caranavi towards La Paz, (15° 48' S, 67° 36' W), 800 m, 20 Oct. 1998, *Krömer & Acebey 136* [SEL!]. Prov. Nor Yungas: Valle del Río Unduavi, between Santa Rosa and Machacamarca, next to the goldmne, (16° 11' 29" S, 67° 42' 00" W), 1440 m, 4 Sept. 1987, *Seidel & Vargas 1081* [LPB!]; ibid.: Milluguaya, 1300 m, Dec. 1917. *Buchtien 4285* [US!]; ibid.: 24 km from Chuspipata towards Yolosa, 16° 17' S, 67° 48' W, 1350 m, 26 Sept. 1989, *Krömer & Acebey 892* [FR!, LPB!, SEL!]; ibid.: 1.5 km from Yolosa on the road to Choroico, 16° 13' S, 67° 44' W, 9 Mar. 1984, *Varadarajan 1285* [MO!]; ibid.: 4.5 km below Yolosa, then 10 km W on road upstream Río Huarinilla, 16° 12' S, 67° 50' W, 1450 m, 19 Oct. 1982, *Solomon 8561* [LPB!, MO!, NY!, SEL!, U!]; ibid.: 14 km from Corocco to Caranavi, 16° 08' 14" S, 67° 42' 25" W, 1141 m, 20 Oct. 2006, *Peters 06.0091* [FR!]; ibid.: 22 km from Coripata to Chulumani, Puente Villa, Puenta de Coripata, (16° 18' S, 67° 30' W), 1400 m, 29 Apr. 1984, *Kinnach 2604* [SEL!; MSBG]; ibid.: 49 km from Coroico to Caranavi, (16° 12' S, 67° 42' W), 880 m, 22 Mar. 1986, *Besse s.n.* [SEL!; MSBG]; ibid.: Between Viscachani and Coroico, 16° 13' 31" S, 67° 45' 12" W, 1420 m, 24 Oct. 2002, *Rex & Schulte 241002-13* [FR!; FAN!]; ibid.: Between Polo Polo and Coroico, 16° 14' 12" S, 67° 45' 36" W, 1380 m, 25 Oct. 2002, *Rex & Schulte 251002-3* [FR!; FAN!]; ibid.: Between Polo Polo and Coroico, 16° 14' 12" S, 67° 45' 36" W, 1380 m, 25 Oct. 2002, *Rex & Schulte 251002-7* [FR!; FAN!]; ibid.: 4 km from Caranavi towards Sapecho, 15° 47' 21"S, 67° 30' 13"W, 917 m, 26 Oct. 2002, *Rex & Schulte 261002-4* [FR!, LPB!; FAN! BGHD!]; ibid.: 15 km from Coroico towards Caranavi, next to Río Coroico, 16° 06' 00" S, 67° 42' 27" W, 1030 m, 26 Oct. 2002, *Rex & Schulte 261002-26* [LPB!; FAN!]; ibid.: Between Yolosa and Caranavi, 16° 13' S, 67° 45' W, 960 m, 13 Feb. 2000, *Vásquez & Gerlach 5654* [USZ!; FAN!]; ibid.: Between Chuspipata and Yolosa, 16° 13' 46" S, 67° 44' 59" W, 1870 m, 24 Oct. 2002, *Vásquez 4642* [SEL! FAN!]; ibid.: 14 km from Chuspipata towards Yolosa, next to Río Huarinillas, 16° 12' S, 67° 45' W, 1150 m, 6 August 1988, *Beck 13880* [LPB!, SEL!]; ibid.: Puente de Coripata, Río Unduavi, 1150 m, 16° 18' S, 67° 30' W, *s.n.* [B 32276!, OSBU!, SEL!, WU 10296!, 9544!; HBV B98B92-1 ex BGHD 102964! ex BGB 148-34-92-40! ex Huntington, USA; BGOS 94-17-0049-80! ex BGM 1992/0643 ex Huntington, USA]. Prov. Sud Yungas: Union of Río Solamaca and Río Tamanpayo, towards Asunta Choquechalca, 16° 06' S, 67° 12' W, 1300 m, 16 Dec. 1997, *Beck 23304* [LPB! (st.), USZ!]; ibid.: Next to Río La Paz, 16° 32' 24" S, 67° 22' 48" W, 1353 m, 24 Feb. 2000, *Nowicki & Müller 2061* [FAN!]; ibid.: Road from Puente Villa to Chulumani, along the Río Unduavi, 16° 24' 05" S, 67° 38' 22" W, 1244 m, 18 Oct. 2006, *Peters 06.0070* [LPB!]; ibid.: Road from Puente Villa to Chulumani, along the Río Unduavi, 16° 24' 05" S, 67° 38' 22" W, 1244 m, 18 Oct. 2006, *Peters 06.0071* [SEL! (st.)]; ibid.: Road from Puente Villa to Chulumani, along the Río Unduavi, 16° 24' 05" S, 67° 38' 22" W, 1244 m, 18 Oct. 2006, *Peters 06.0072* [FR!]; ibid.: Road from Puente Villa to Chulumani, along the Río Unduavi, 16° 24' 05" S, 67° 38' 22" W, 1244 m, 18 Oct. 2006, *Peters 06.0073* [LPB! (st.)]; ibid.: 6 km on the road from Chulumani to Tajma, along the Río Unduavi, next to Puente Villa, 16° 23' 38" S, 67° 29' 56" W, 1541 m, 19 Oct. 2006, *Peters 06.0074* [LPB!; BGHD!]; ibid.: 6 km on the road from Chulumani to Tajma, along the Río Unduavi, next to Puente Villa, 16° 23' 38" S, 67° 29' 56" W, 1541 m, 19 Oct. 2006, *Peters 06.0075* [SEL!]; ibid.: 6 km on the road from Chulumani to Tajma, along the Río Unduavi, next to Puente Villa, 16° 23' 38" S, 67° 29' 56" W, 1541 m, 19 Oct. 2006, *Peters 06.0076* [BGHD!]; ibid.: 18 km on the road from Chulumani to Choquechaca, along the Río Unduavi, behind Tajma, 26° 21' 42" S, 67° 27' 48" W, 1503 m, 19 Oct. 2006, *Peters 06.0077* [FR! (st.)]; ibid.: 18 km on the road from Chulumani to Choquechaca, along the Río Unduavi, behind Tajma, 26° 21' 42" S, 67° 27' 48" W, 1503 m, 19 Oct. 2006, *Peters 06.0078* [SEL! (st.); BGHD!]; ibid.: 32 km on the road from Chulumani to Choquechaca, along the Río Unduavi, behind Tajma, 16° 17' 42" S, 67° 23' 43" W, 1091 m, 19 Oct. 2006, *Peters 06.0082* [LPB!]; ibid.: Road from Chulumani to La Asunta, behind Choquechaca, next to Río Boopi, 16° 09' 14" S, 67° 10' 47" W, 730 m, 19 Oct. 2006, *Peters 06.0085* [SEL!; BGHD!]; ibid.: 93 km on the road from Chulumani to La Asunta, behind Choquechaca, next to Río Boopi, 16° 05' 20" S, 67° 10' 52" W, 781 m, 19 Oct. 2006, *Peters 06.0086* [FR!; BGHD!]; ibid.: 16 km on the road from Chulumani to Puente Villa, 16° 23' 10" S, 67° 36' 13" W, 1375 m, 20 Oct. 2006, *Peters 06.0089* [LPB!]; ibid.: 16 km on the road from Chulumani to Puente Villa, 16° 23' 10" S, 67° 36' 13" W, 1375 m, 20 Oct. 2006, *Peters 06.0090* [SEL!; BGHD!]; ibid.: 9 km from Chamaca towards Chulumani, Río La Paz, 16° 17' S, 67° 17' W, 900 m, 3 Oct. 1995, *Kessler 5726* [SEL! (fr.)]; ibid.: Between Irupana and Yuni, 16° 27' S, 67° 26' W, 1600 m, 12 Feb. 2002, *Vásquez 3633* [FR!, LPB!, SEL!, USZ!, VASQ!, WU!]; ibid.: 50 km from Chulumani towards La Asunta, 16° 11' 29" S, 67° 11' 00" W, 890 m, 30 May 1986, *Beck 12612* [LPB!, SEL!]; ibid. 26 km from Chulumani towards Asunta, behind Tajma, 16° 18' 00" S, 67° 19' 12" W, 1300 m, 27 June 1985, *Beck 12076* [LPB!, SEL!, USZ!]; ibid.: 5 km from Puente Villa towards Unduavi, 16° 19' S, 67° 50' W, 1470 m, 1 July 1981, *Beck 4701* [LPB!, US!]; ibid.: Río Unduavi, (16° 18' S, 67° 24' W), 1000 m, 16 Aug. 1976, *Rauh 40583* [FR! (st.), HEID!, SEL! (st.), WU!]. Prov. Inquisivi: Between Samaipata and Charrupampa, 22 km behind Choquetanga, Comunidad Khora, 16° 40' S, 67° 20' W, 1420 m, 29 Oct. 1994, *Salinas 2968* [LPB! (st.)]; ibid.: 15 km from Circuata to Chulumani, (16° 30' S, 67° 24' W), 1370 m, 30 Jan. 1983, *Besse 1845* [SEL!; MSBG]; ibid.: Parque Nacional de Choquecamiri, 20 km N of Choquetanga, mouth of Río Aguilani, Lakachaka-Camp, following Río Miguillas about 1km upstream from the footbridge, 16° 40' S, 67° 20' W, 1450 m, 19 Sept. 1991, *Lewis 40358* [LPB!, MO!]. **Dpto. Cochabamba:** Prov. Ayopaya: Atispaya, 16° 36' S, 66° 43' W, 1330 m, 22 June 2001, *Vásquez 4177 a* [LPB!].

WITHOUT LOCALITY: 26 Mar. 1986, *Rauh 38324 a* [HEID!], s.n. [WU 10238!, 10227!; HBV B01B83], s.n. [SEL 090166!; MSBG s.n.], s.n. [SEL 090167!; MSBG s.n.], s.n. [HBG 31-1-91!; BGHH 81-G-0496! ex living collection Hemker].

Distribution and ecology

Range size: Small.

Countries: BOLIVIA. Dpto. La Paz, Cochabamba.

Ecoregions: Sub-Andean-Amazon-Forests, Yungas, Inter-Andean-Dry-Forests.

Life style & habitats: Mainly humid, evergreen lowland and montane forests, but in dry, deciduous valleys as well. Terrestrial or saxicolous in more or less disturbed montane forests. Steep slopes along the road, exposed and dry to shady and moist understory. Semideciduous forest with *Anadenanthera colubrina* (Fabaceae), *Astronium urundeuva* (Anacardiaceae), *Aspidosperma cylindrocarpum* (Apocynaceae), *Oxandra espintana* (Annonaceae), *Acacia loretensis* (Fabaceae), *Erythroxylum rotundifolium* (Erythroxylaceae) and *Ceiba* sp. (Bombaceae) (*Uzquiano 207* [LPB]). Semideciduous forest with *Anadenanthera colubrina* (Fabaceae), *Trichilia elegans* (Meliaceae), *Gynerium sagittatum* (Poaceae) and *Cecropia* sp. (Cecropiaceae) (*Araujo*

2278 [LPB]). Sub-Andean-Amazon-Forests with *Juglans boliviana* (Juglandaceae) and *Cariniana estrellensis* (Lecythidaceae) (*Fuentes 9396* [LPB]). Rocky slope with *Crassula* sp. (Crassulaceae), *Encyclia* sp. (Orchidaceae) and *Cleistocactus* sp. (Cactaceae) (*Nowicki & Deichmann 2356* [FR]).

Altitude: 700–1900 m.

Taxonomic delimitation and systematic relationships

Fosterella rusbyi is very similar to *F. vasquezii* but differs in the considerably larger inflorescence, the wider leaf blades which are clearly serrate and undulate and the longer peduncle bracts and petals. According to molecular data (based on AFLPs as well as on combined sequence data of four chloroplast loci), *F. rusbyi* and *F. vasquezii* are closely related, but clearly distinct (REX et al. 2009).

Fosterella elata was recombined to synonymy under *F. rusbyi* in PETERS et al. (2008 b) for following reasons: *Fosterella rusbyi* was only known from a few collections made in the first half of the last century, when LUTHER (1981) described *F. elata*. He delimited it from *F. penduliflora* but did not compare it to *F. rusbyi*. Maybe he did not take into account *F. rusbyi* as a similar species, because SMITH & DOWNS (1974) stated that it had greenish petals. This was a shortening of the original description by Mez (1901) who wrote: "Petals twice as long as the sepals, apparently greenish". Since the original description, *F. elata* has frequently been collected and turned out to be a rather common species in the understory of the Yungas rain forests of the La Paz Department. After careful revision of the type material of *F. rusbyi* it became clear that it is impossible to detect any differences between *F. rusbyi* and *F. elata*. The latter name therefore was reduced to synonymy.

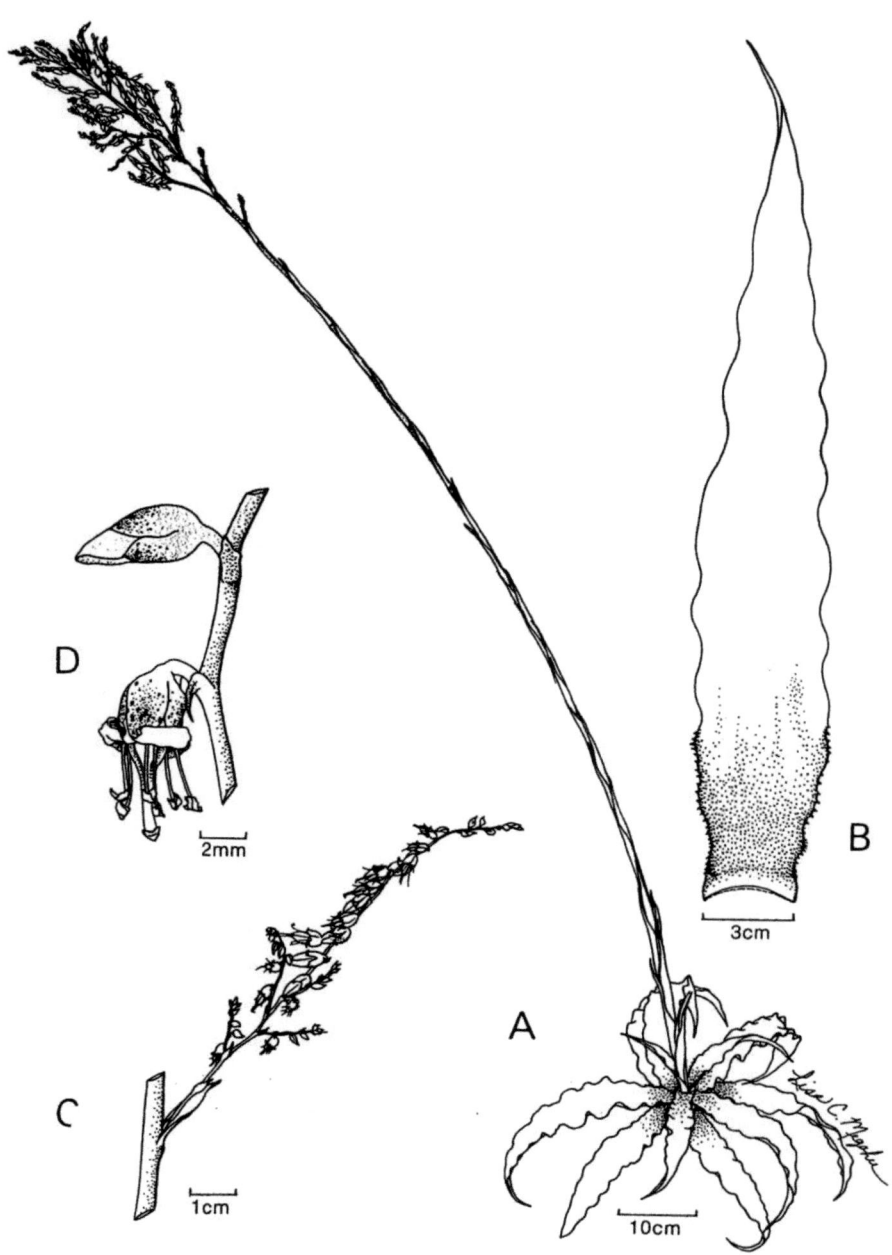

Fig. 100: *Fosterella rusbyi*. A Habit. B Leaf. C Inflorescence branch. D Flower.
Drawing of *Luer et al. s.n.* by L.C. McGahee in LUTHER (1981). © Selbyana.

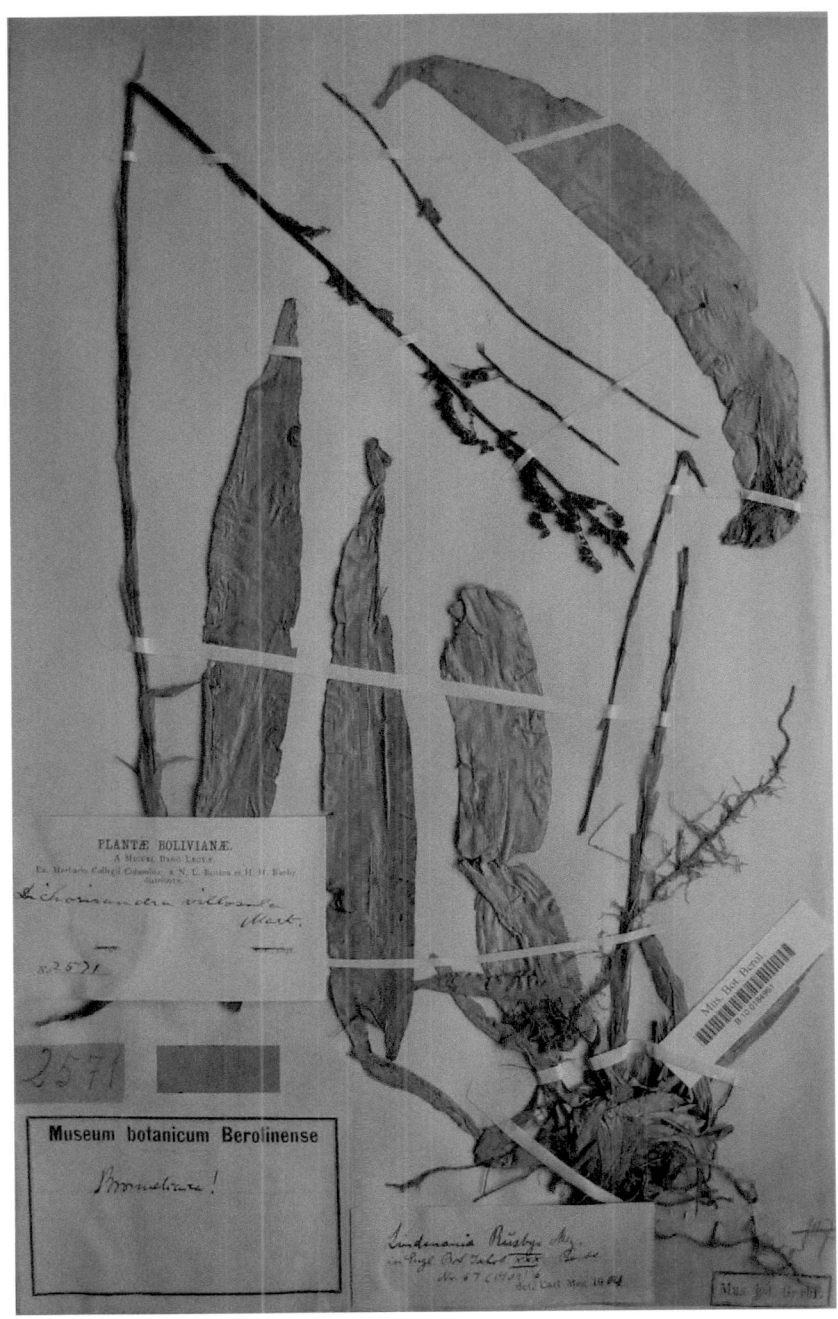

Fig. 101: Lectotype of *Fosterella rusbyi* (Mez) L.B. Sm. [B].

Fig. 102: *F. rusbyi (Peters 06.0074)* in the natural habitat.

Fig. 103: Flowers of *F. rusbyi (Peters 06.0078)*.

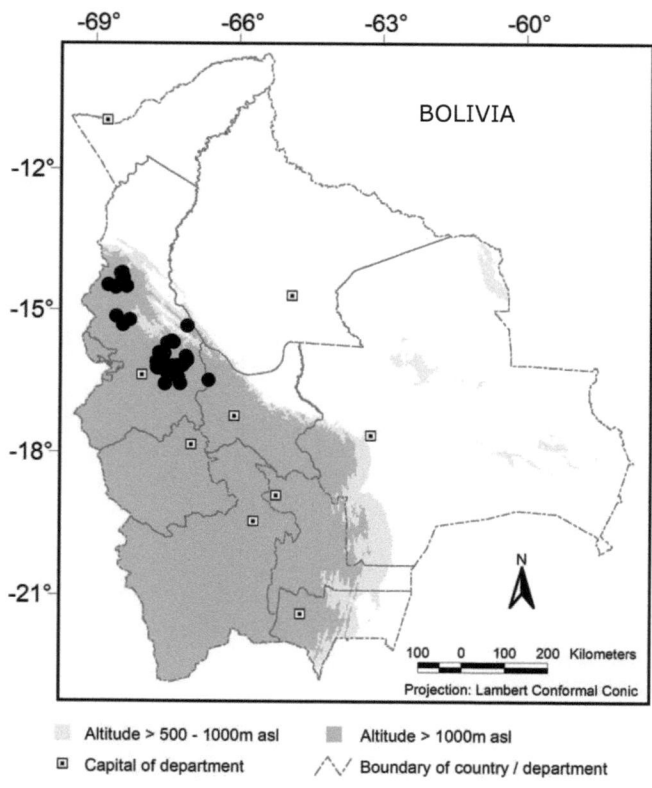

Fig. 104: Distribution of *Fosterella rusbyi*.

Fosterella schidosperma (Baker) L.B. Sm., Phytologia 8: 500. 1963. Fig. 105 & 106.

Basionym:	*Schidospermum sanseviera* Griseb ex Lechl., Berberid. Amer. Austr. 56. 1857. Nomen nudum based on *Lechler 2382*.
≡	*Chlorophytum schidospermum* Baker, J. Linn. Soc., Bot. 15: 326. 1876. TYPE: Perú. Shady woods next to San Govan, July 1854, *Lechler 2382* [Lectotype: G (phot!)].
=	*Cottendorfia rusbyi* Baker ex Rusby, Bull. Torrey Bot. Club 29: 697.1902. TYPE: Bolivia. Yungas, growing upon rocky, partly shaded banks, 2000 m, 1885, *Rusby 2541* [Lectotype: NY!, designated here; Isotypes: US!, F!].

Comment on type

As SMITH & DOWNS (1974) already stated, the correct name of the type locality is "San Gabán" and not "San Govan" as given in the original description. "San Gabán is located in the Province of Carabaya in the Department of Puno at approx. 13° 20' S, 70° 20' W, 1000 m. "San Govan" could neither be found on any map, nor did local people know the name.

The type material (collection *Lechler 2382*) caused a lot of confusion. The description of *Chlorophytum schidospermum* Baker is based on three specimens [two from G and one from K] with the same collection number, but actually they are not the same species. Following the ICBN, Art. 9.12 and Recommendation 9A.5 (MCNEILL et al. 2005), dealing with heterogeneous type material, the name must remain attached to that part which corresponds most closely to the original description or diagnosis. In this case, this is the specimen *Lechler 2382* [G 00074337]) which therefore is designated as lectotype for *Fosterella schidosperma* (Baker) L.B. Sm. here. The other two specimens, *Lechler 2382* [G 00074338 and K 000321537] reveal clear accordance with *Weberbauer 5635* and have to be identified as *Fosterella weberbaueri* (Mez) L.B. Sm., that was described as *Lindmania weberbaueri* Mez in 1913. The type material of *Cottendorfia rusbyi* Baker, which is combined to synonymy under *Fosterella schidosperma* (Baker) L.B. Sm. here, represent most beneficial complement to the lectotype of *F. schidosperma*.

As BAKER ex RUSBY (1902), when he described *Cottendorfia rusbyi*, did not choose a holotype from among the three specimens on hand, the specimen preserved in NY is designated as lectotype here. The differentiation of *Fosterella schidosperma* (Baker) L.B. Sm. var. *vestita* L.B. Sm. & Read is abolished here. Both specimen cited in the original description, *Dillon 1253* [US!, F!, MO!] and *Dillon 1201* [F!, US!] have to be identified as *Fosterella weberbaueri* according to the concept explicated above. In *F. weberbaueri*, a slight indumentum of particularly young inflorescence is quite common. Therefore, *Fosterella schidosperma* (Baker) L.B. Sm. var. *vestita* L.B. Sm. & Read is reduced to synonymy under *Fosterella weberbaueri* (Mez) L.B. Sm. here.

Etymology

The epithet refers to the basionym, the monotypic *Schidospermum* Griseb. (nomen nudum).

Description

Plant acaulescent, up to 80 cm high. **Leaves** few, up to 7, forming an open, more or less upright rosette. **Sheaths** 20–30 mm wide, entire, whitish, glabrous. **Blades** narrowly lanceolate, acuminate, narrowed towards the base, 35–40 cm long and 2–3 cm wide, thin, entire, adaxially glabrous, abaxially scattered lepidote by peltate trichomes with dentate margin, glabrescent. **Peduncle** 25–40 cm long, to 5 mm in diameter, green, glabrous. **Peduncle bracts** 1.5–4 cm long, shorter than or equalling the internodes, entire, abaxially scattered lepidote, glabrescent. **Inflorescence** paniculate with branches up to 2nd order, 40–45 cm long and 10–25 cm wide, axes green, glabrous. **Primary bracts** 10–15 mm long, longer than or equalling the sterile base of the branches, entire, abaxially glabrous. **Branches** 15–20, ascending, straight, 10–15 cm long, bearing up to 40 flowers. **Secondary branches** 5–15 cm long, bearing up to 25 flowers. **Floral bracts** 2 mm long, shorter than the pedicels, entire, abaxially glabrous. **Flowers** subsecund, pendulous, 4–6 mm apart. **Pedicels** 2–3 mm long. **Sepals** 2–3 mm long, green, glabrous. **Petals** 5–6 mm long, white, recoiled like watchsprings during anthesis and afterwards. **Filaments** 3 mm long. **Anthers** 2.5 mm long. **Style** 2.5 mm long. **Stigmatic complex** simple-erect. **Capsule** ovoid, 4 mm long, 2.5 mm wide. **Seeds** filiform, bicaudate, 2 mm long.

Specimens seen

PERÚ: **Dpto. Junín:** Prov. Chanchamayo: Chanchamayo Valley, (10° 51' S, 75° 03' W), 1500 m, 1924–1927, *Schunke 273* [F! (st.)]; ibid.: Río Perené, south of Santa. Ana, 11° 22' 13" S, 74° 44' 10" W, 700 m, 18 Mar. 1997, *Hromadnik 23177 a* [WU!; HBV B97B129-1]; ibid.: Río Paucartambo Valley, near Perené Bridge, (10° 54' S, 75° 54' W), 700 m, 19 June 1929, *Killip & Smith 25326* [NY!, US!]; ibid.: La Merced, (11° 00' S, 75° 18' W), 650 m, 10–24 Aug. 1923, *Macbride 5352* [F!, GH!]. Prov. Satipo: Santa Ana, between Mazamari and San Martin de Pongoa, 11° 24' S, 74° 36' W, 650 m, 16 Sept. 1981, *Fernández 7* [WU!; MSBG].
BOLIVIA: **Dpto. La Paz:** Prov. Caranavi: 19 km from Caranavi towards Palos Blancos, 500 m downstream from bridge across Río Carrasco, 15° 45' S, 67° 29' W, 860 m, 29 Sept. 2001, *Jiminez & Bach 661* [LPB! (fr.), SEL! (fr.)]; ibid.: 17 km from Caranavi towards Palos Blancos, 15° 48' S, 67° 30' W, 1020 m, 29 Sept. 2001, *Jiminez & Bach 681* [SEL! (fr.)].

Distribution and ecology

Range size:	Small.
Countries:	PERÚ. Dpto. Junín, Puno. BOLIVIA. Dpto. La Paz.
Ecoregions:	Peruvian and Bolivian Yungas, Pre-Andean-Forests, Inter-Andean-Dry-Forests.
Life style & habitats:	Humid, evergreen lowland and montane forests and dry, deciduous valleys. Rocky, shaded riverbanks and steep slopes. Shrubberies and more or less open forests. Deciduous forest in dry interandean valley with *Anadenanthera colubrina* (Fabaceae) (*Jiminez & Bach 661* [LPB]).
Altitude:	650–2000 m.

Taxonomic delimitation and systematic relationships

As IBISCH et al. (2002) already elaborated when they reinstated *Fosterella weberbaueri*, *F. schidosperma* and *F. weberbaueri* are strikingly similar. The only clear-cut character for destinction is the shape of the petals, which are recoiled like watchsprings and stay so poastanthesis in *F. schidosperma*, whereas in *F. weberbaueri*, they are only more or less recoiled and straight postanthesis. Furthermore, the leaf blades in *F. schidosperma* tend to be long and narrow, while in *F. weberbaueri* they are rather short and wide. The inflorescence of *F. schidosperma* all in all is slightly larger and more ramified, the flowers are subsecund and pendulous, whereas they are spreading and suberect in *F. weberbaueri*. Hitherto, no molecular data for *F. schidosperma* is available.

Fig. 105: Lectotype of *Fosterella schidosperma* (Baker) L.B. Sm. [G].
© Conservatoire et Jardin botaniques de la Ville de Genève.

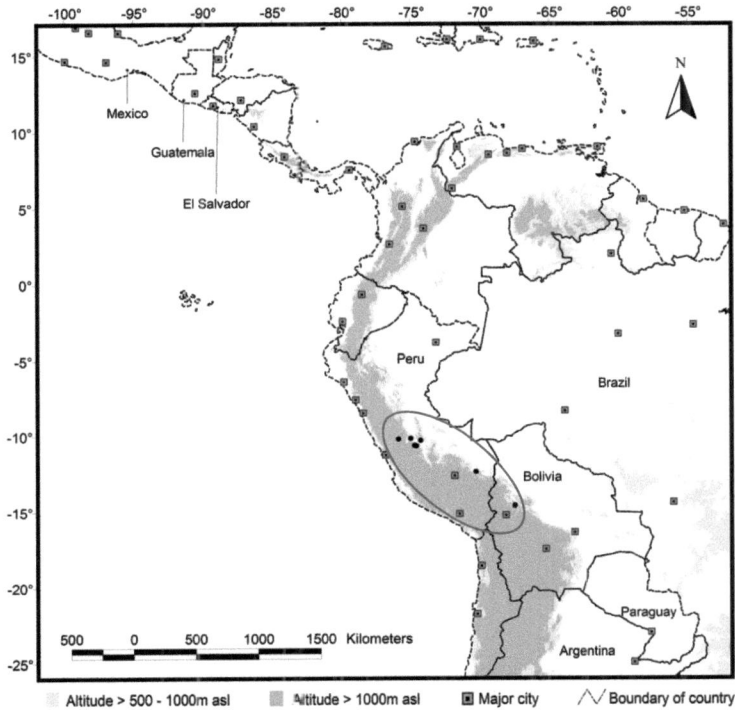

Fig. 106: Distribution of *Fosterella schidosperma*.

Fosterella spectabilis H. Luther, J. Bromeliad Soc. 47 (3): 118. 1997. Fig. 107–111.

> TYPE: Bolivia. Dpto. Santa Cruz: Near Angostura, 1700 m, Aug. 1993, *D. Cathcart B-17*; flowered in greenhouse MSBG 1995-0415, 12 Mar. 1996, *H. Luther s.n.* [Holotype: SEL!; Isotype: HB!].

Comment on type

Angostura is located in the Province of Florida at approx. 18° 10' S, 63° 41' W. However, the species was never confirmed here. Clonotypes are cultivated in: Marie Selby Botanical Gardens, Sarasota (Florida); private living collection of Elton Leme, Teresópolis (Brazil) ex MSBG; Botanical Garden Berlin ex Botanical Garden Vienna (Austria) ex Deroose Nursery ex MSBG.

Etymology

The epithet refers to the relatively spectacular (for a *Fosterella*) long and bright red flowers.

Description

Plant acaulescent, up to 90 cm high. **Leaves** few, 10–15, forming an open, arched rosette. **Sheaths** 35–60 mm wide, entire, whitish to tinged and spottet reddish, glabrous. **Blades** narrowly oblanceolate, acuminate, narrowed towards the base, 25–70 cm long and 3.5–6 cm wide, succulent towards the base, entire, adaxially glabrous, abaxially tinged and spottet reddish towards the base, scattered lepidote by peltate trichomes with dentate margin, glabrescent. **Peduncle** 15–40 cm long, to 5 mm in diameter, green, glabrous, glaucous. **Peduncle bracts** 2 cm long, more or less equalling the internodes, entire, abaxially scattered lepidote, glabrescent. **Inflorescence** racemose or compound racemose, 15–55 cm long and to 15–30 cm wide, axes green, glabrous, glaucous. **Primary bracts** 20 mm long, shorter than or equalling the sterile base of the branches, entire, abaxially scattered lepidote, glabrescent. **Branches** up to 10, subsecund, inclined, arcuate, 10–30 cm long, bearing up to 30 flowers. **Floral bracts** 8–12 mm long, shorter than the pedicels, entire, reddish, abaxially glabrous. **Flowers** secund, pendulous, 5–15 mm apart. **Pedicels** 8–15 mm long. **Sepals** 8–9 mm long, pale green to reddish, glabrous, glaucous. **Petals** 15–24 mm long, reddish to coral-orange, straight during anthesis and afterwards. **Filaments** 12–15 mm long, the inner filaments adnate to the petals for 3 mm from their base. **Anthers** 2 mm long. **Style** 13–15 mm long. **Stigmatic complex** simple-erect. **Capsule** ovoid, 7–8 mm long, 3–4 mm wide. **Seeds** filiform, bicaudate, 2–3 mm long.

Specimens seen

BOLIVIA: **Dpto. Santa Cruz:** Prov. Florida: Near Angostura, 18° 10' S, 63° 41' W, 1700 m, Aug. 1993, *D. Cathcart B-17*; flowered in MSBG 1995-0415, 22 Feb. 1995, *D. Cathcart s.n.* [SEL!]. **Dpto. Chuquisaca:** Prov. Jaime Mendoza: 68 km from Monteagudo towards Padilla, 19° 34' S, 64° 09' W, 1400 m, 22 July 1995, *Kessler 5028* [SEL!]; ibid.: 53 km from Padilla towards Monteagudo, Río Mojo Torillo, 19° 32' 18" S, 64° 09' 02" W, 1300 m, 6 Dec. 2004, *Vásquez*

4955 [VASQ!]. Prov. Hernando Siles: 64 km from Monteagudo to Padilla, 19° 32' 19" S, 64° 07' 28", 1269 m, 4 Oct. 2006, *Peters 06.0045* [LPB!; BGHD!]; bid.: 64 km from Monteagudo to Padilla, 19° 32' 19" S, 64° 07' 28", 1269 m, 4 Oct. 2006, *Peters 06.0046* [LPB!; BGHD!]; ibid.: 64 km from Monteagudo to Padilla, 19° 32' 19" S, 64° 07' 28", 1269 m, 4 Oct. 2006, *Peters 06.0047* [SEL!; BGHD!]; ibic.: 64 km from Monteagudo to Padilla, 19° 32' 19" S, 64° 07' 28", 1269 m, 4 Oct. 2006, *Peters 06.0048* [FR!; BGHD!] Prov. Sud Cinti: Serranía La Polla, 20° 45' 11" S, 64° 13' 59" W, 930 m, *Serrano 4955* [HSB!].

Distribution and ecology

Range size:	Small.
Countries:	BOLIVIA. Dpto. Santa Cruz, Chuquisaca.
Ecoregion:	Tucuman-Bolivian-Forests.
Life style & habitats:	More or less open, semihumid, evergreen to semideciduous forests. Small ravines in evergreen forest (*Kessler 5028* [SEL]).
Altitude:	900–1700 m.

Taxonomic delimitation and systematic relationships

Fosterella spectabilis differs from all other species of the genus, in its long, reddish to coral-orange flowers with a semi-tubular corolla. Since this goes along with abundant nectar production, it can be assumed that *F. spectabilis* is an ornithophilous species (the only one in the genus).

According to molecular data, *F. spectabilis* belongs to one clade together with *F. floridensis*, *F. yuvinkae*, *F. windischii*, *F. vasquezii* and *F. rusbyi*. Two accessions of *F. spectabilis* form a well-supported subclade which is sister to the remainder of the group.

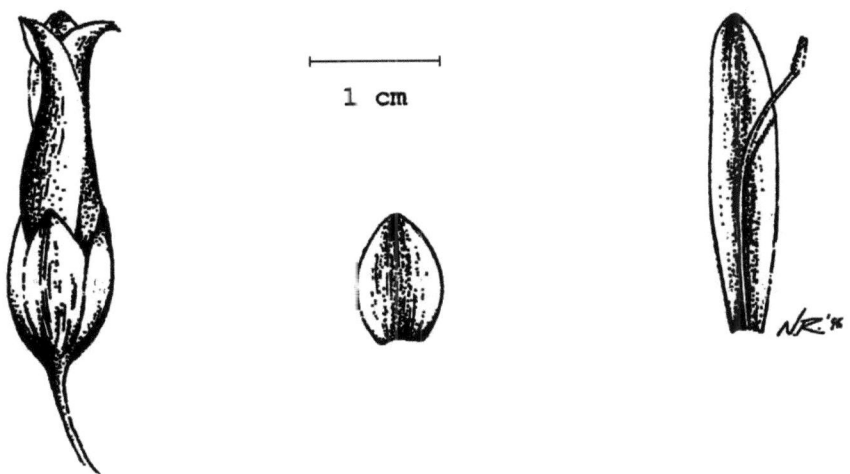

Fig. 107: *Fosterella spectabilis*. A Flower at anthesis. B Sepal. C Petal with stamen. Drawing of *Luther s.n.* by N. Rosseland in LUTHER (1997). © Selbyana.

Fig. 108: Holotype of *Fosterella spectabilis* H. Luther [SEL].

Fig. 109: *F. spectabilis (Peters 06.C045).*

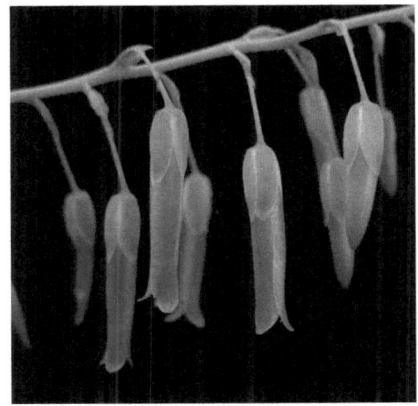

Fig. 110: Flowers of *F. spectabilis (Peters 06.0045).*

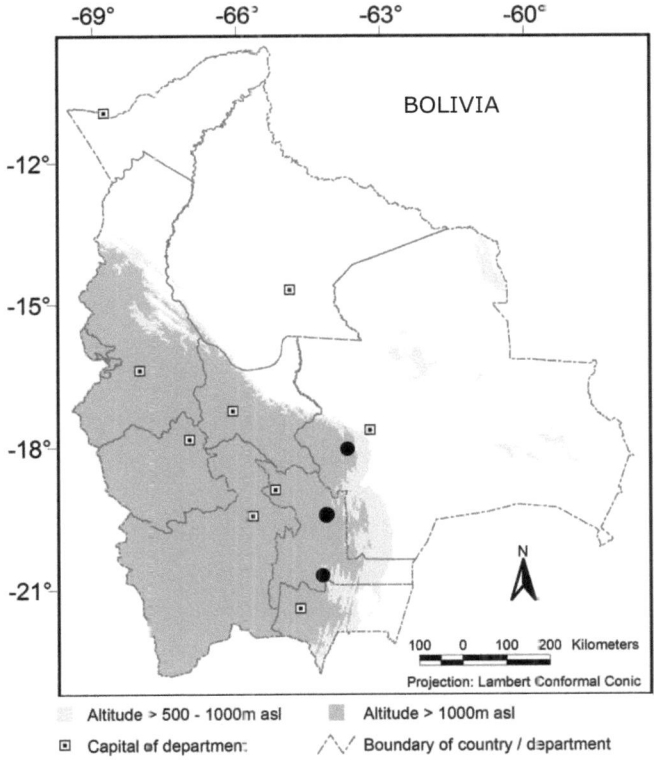

Fig. 111: Distribution of *Fosterella spectabilis.*

Fosterella vasquezii E. Gross & Ibisch, J. Bromeliad Soc. 47 (5): 212. 1997. Fig. 112–115.

TYPE: Bolivia. Dpto. Santa Cruz: Prov. Velasco, Parque Nacional "Kempff Mercado", growing on steep sandstone rocks of the Serranía de Huanchaca, a mountain chain on the Bolivian-Brazilian border, nearby "El Encanto" waterfall, 200 m, 27 Aug. 1993, *Ibisch et al. 93.0652* [Holotype: HEID!; Isotype: LPB!].

Comment on type

According to the original description (IBISCH et al. 1997), the type locality is "a site characterised by rather humid conditions with water in all probability permanently running down the stone wall on which, among others, a species of *Urtricularia* grows. The surrounding forest is humid and semi-evergreen. Among typical tree species to be found is *Swietenia macrophylla*". It is located at approx. 14° 40' S, 60° 30' W.

Clonotypes are cultivated in the Botanical Gardens Berlin, Bonn, Heidelberg (all Germany) and Vienna (Austria).

Etymology

This species is dedicated to Roberto Vásquez (*1941), specialist of several Bolivian plant groups, particularly Orchidaceae and Bromeliaceae.

Description

Plant acaulescent, up to 50 cm high. **Leaves** few, 10–15, forming an open, flat rosette. **Sheaths** to 50 mm wide, obscurely serrate, whitish, glabrous. **Blades** lanceolate, 25–45 cm long and 1.5–3.5 cm wide, succulent and obscurely serrate towards the base, undulate, adaxially glabrous, abaxially covered by a thick layer of interwoven, peltate trichomes with fimbriate margin. **Peduncle** 20–30 cm long, 4 mm in diameter, green, glabrous. **Peduncle bracts** 3 cm long, longer than the internodes, entire, abaxially covered by a layer of interwoven trichomes. **Inflorescence** paniculate with branches up to 2nd order, 30 cm long and 15 cm wide, axes green, glabrous. **Primary bracts** 10 mm long, shorter than the sterile base of the branches, entire, abaxially interwoven lepidote. **Branches** 10–20, ascending, straight, 10–15 cm long, bearing 10–25 flowers. **Secondary branches** 3–5 cm long, bearing up to 10 flowers. **Floral bracts** 2–3 mm long, longer than or equalling the pedicels, entire, abaxially glabrous. **Flowers** secund, pendulous, 4–6 mm apart. **Pedicels** 2–3 mm long. **Sepals** 2 mm long, green, glabrous. **Petals** 5 mm long, white, recoiled like watchsprings during anthesis and afterwards. **Filaments** 4 mm long. **Anthers** 2 mm long. **Style** 3 mm long. **Stigmatic complex** simple-erect. **Capsule** ovoid, 5–6 mm long, 3 mm wide. **Seeds** filiform, bicaudate, 2 mm long.

Specimens seen

BOLIVIA: **Dpto. Santa Cruz:** Prov. Velasco: Parque Nacional Noel Kempff, Los Fierros, "El Encanto" Waterfall, 14° 45' S, 60° 35' W, 300 m, 5 Aug. 1998, *Ibisch 98.0116* [FAN!]; ibid.: Parque Nacional Noel Kempff, Las Gamas, 14° 48' S, 60° 26' W, 700 m, 13 June 1994, *Vásquez 2554* [LPB!, USZ!]; ibid.: Parque Nacional Noel Kempff, from Senda de Los Fierros towards Sierra de Huanchaca, 8 km from the end of the track, 14° 33' 16" S, 60° 47' 48" W, 300 m, 13 June 1994, *Killeen 6469* [SEL!]; ibid.: Parque Nacional Noel Kempff, Las Gamas, 14° 48' 30" S, 60° 24' 17" W, 850 m, 28 Oct. 1995, *Killeen & Grinwood 7800* [MO!, SEL!, USZ!].

Distribution and ecology

Range size:	Very small.
Country:	BOLIVIA. Dpto. Santa Cruz.
Ecoregion:	Beni-and-Santa-Cruz-Amazon-Forests.
Life style & habitats:	Southwest Amazon forests in transition to Chiquitano-Dry-Forest; humid habitats associated with waterfalls of the Caparús table mountain (Precambrian rocks of the Brazilian shield).
Altitude:	200–850 m.

Taxonomic delimitation and systematic relationships

Fosterella vasquezii is similar to *F. rusbyi* but differs in narrower leaf-blades, which are only obscurely serrate and not at all narrowed towards the base. The inflorescence all in all is smaller and particularly the peduncle bracts and petals are shorter.

According to the molecular trees, both based on AFLP and chloroplast-sequence data, *F. vasquezii* and *F. rusbyi* are closely related, but clearly separated. The two belong to one clade along with *F. windischii*, *F. yuvinkae*, *F. floridensis* and *F. spectabilis*.

Fig. 112: Holotype of *Fosterella vasquezii* E. Gross & Ibisch [HEID].

Fig. 113: *F. vasquezii (Ibisch 93.0652)* in the natural habitat. Photo: P. Ibisch.

Fig. 114: *F. vasquezii (Ibisch 93.0652)*.

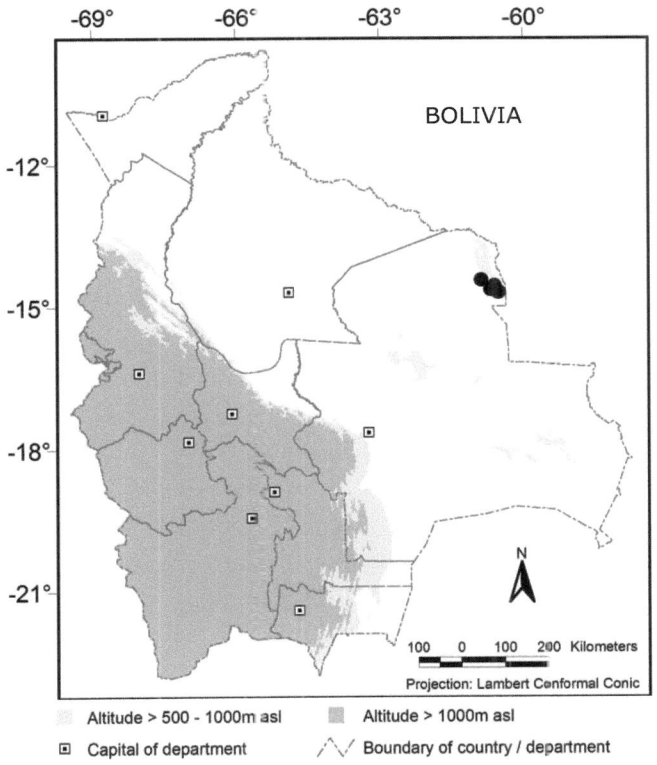

Fig. 115: Distribution of *Fosterella vasquezii*.

Fosterella villosula (Harms) L.B. Sm. Phytologia 7: 172. 1960. Fig. 116–119.

Basionym: *Lindmania villosula* Harms, Notizblatt 10: 794. 1929. TYPE: Bolivia. Dpto. Cochabamba: Between Incachaca and San Antonio, terrestrial, in the wood, 1600 m, July 1926, E. *Werdermann 2120* [Lectotype: B!; Isotype: MO!].

Comment on type

The type locality is situated in the Province Chapare, probably at approx. 17° 14' S, 65° 49' W. When HARMS (1929), based on the collection *Werdermann 2120*, described *Lindmania villosula*, he did not choose a holotype from among the two specimens. He made some notes on the voucher preserved in B, but did not explicitly marked it as holotype. Therefore, in accordance with the ICBN (2006), the designation of the specimen preserved in B by SMITH & DOWNS (1974) as holotype should be changed to lectotype.

Etymology

The epithet refers to the strongly villous inflorescence.

Description

Plant acaulescent, up to 100 cm high. **Leaves** many, 10–15, forming a dense, arched rosette. **Sheaths** to 25 mm wide, entire, whitish, glabrous. **Blades** lanceolate, acuminate, narrowed towards the base, 35–60 cm long and 2.5–6 cm wide, thin, entire, adaxially glabrous, abaxially frequently reddish, scattered lepidote by peltate trichomes with dentate margin. **Peduncle** 35–70 cm long, 3–5 mm in diameter, green, densely villous-arachnoid. **Peduncle bracts** 3–6 cm long, longer than the internodes, entire, abaxially scattered lepidote. **Inflorescence** paniculate with branches up to 2nd order, 15–35 cm long and 5–20 cm wide, axes green, densely villous. **Primary bracts** 15–20 mm long, shorter than the sterile base of the branches, entire, abaxially villous. **Branches** 10–15, inclined, arcuate, 6–12 cm long, bearing up to 25 flowers. **Secondary branches** 2–5 cm long, bearing up to 10 flowers. **Floral bracts** 7 mm long, longer than the pedicels, entire, abaxially villous. **Flowers** secund, pendulous, subsessile, 3–5 mm apart. **Pedicels** 1–2 mm long. **Sepals** 4–5 mm long, green, sparsely villous. **Petals** 7 mm long, white, sometimes yellowish, recurved during anthesis, straight afterwards. **Filaments** 5 mm long. **Anthers** 2 mm long. **Style** 3–4 mm long. **Stigmatic complex** conduplicate-spiral. **Capsule** ovoid, 5–6 mm long, 3 mm wide. **Seeds** filiform, bicaudate, 2–3 mm long.

Specimens seen

BOLIVIA: Dpto. **La Paz:** Without precise locality, 2002, *Nowicki s.n.* [SEL!]; ibid.: Río Boopi Valley, 1000 m, 22 Aug. 1921, *Rusby 393* [NY!]. Prov. Iturralde: Parque Nacional Madidi, Río Tuichi, Arroyo Rudidi, Pica de Cazadores towards Serranía Eslabón, 14° 21' 58" S, 68° 00' 45" W, 763 m, 4 Oct. 2002, *Fuentes 5366* [LPB!, VASQ!]. Prov. Franz Tamayo: Hills between Río Machariapo and Río Ubito, 14° 25' S, 68° 31' W, 900 m, 22 July 1993, *Kessler 4109* [LPB!,

SEL!]; ibid.: Parque Nacional Madidi, between Carjata and the old cemetery of Mojos, 14° 33' 48" S, 68° 52' 07" W, 1500 m, 1 July 2005, *Fuentes 9161 A* [LPB!]; ibid. Parque Nacional Madidi, between Carjata the old cemetery of Mojos, 14° 33' 48" S, 68° 52' 07" W, 1500 m, 1 July 2005, *Fuentes 9196* [LPB!]; ibid.: Area Natural de Manejo Integrado Madidi; Unapa, 21 km N of Apolo, 14° 32' 26" S, 68° 29' 46" W, 1022 m, 1 Sept. 2004, *Fuentes & Aldana 6374* [LPB!]. **Prov. Muñecas:** Along Río San Cristobal, next to Consata, 1200 m, 22 Jan. 1981, *Besse s.n.* [F!, SEL!; MSBG]. **Prov. Nor Yungas:** 15 km from Coroico towards Caranavi, next to Río Coroico, 16° 06' 00" S, 67° 42' 27" W, 1030 m, 26 Oct. 2002, *Rex & Schulte 261002-14* [LPB!; FAN!]; ibid.: 15 km from Coroico towards Caranavi, next to Río Coroico, 16° 06' 00" S, 67° 42' 27" W, 1030 m, 26 Oct. 2002, *Rex & Schulte 261002-15* [FR!]; ibid.: 20 km from Yolosa to Caranavi, 1050 m, 26 Oct. 2002, 16° 05' 32" S, 67° 42' 15" W, *Vásquez 4650 a* [FR!; FAN!]. **Dpto. Beni: Prov. Ballivian:** 3 km from El Sillar towards Palos Blancos, 15° 17' 08" S, 67° 10' 23" W, 850 m, 27 Oct. 2002, *Rex & Schulte 271002-4* [FAN!]. **Dpto. Cochabamba:** Prov. Chapare: Road from Villa Tunari to Cochabamba, 1 km before Río Tiumany, 17° 04' 12" S, 65° 39' 00" W, 651 m, 17 Oct. 2002, *Rex & Schulte 171002-1* [FR!]; ibid.: Road from Villa Tunari to Cochabamba, between Naranjitos and Corani, 17° 03' 40" S, 65° 38' 40" W, 626 m, 23 Oct. 2006, *Peters 06.0105* [LPB!; BGHD!]; ibid.: Road from Villa Tunari to Cochabamba, between Naranjitos and Corani, 17° 03' 40" S, 65° 38' 40" W, 626 m, 23 Oct. 2006, *Peters 06.0106* [SEL!]; ibid.: Road from Villa Tunari to Cochabamba, between Naranjitos and Corani, 17° 03' 40" S, 65° 38' 40" W, 626 m, 23 Oct. 2006, *Peters 06.0107* [FR!]; ibid.: Road from Villa Tunari to Cochabamba, between Río Huayruruni and Río Huañuska, 17° 01' 02" S, 65° 33' 23" W, 433 m, 23 Oct. 2006, *Peters 06.0108* [LPB!]; ibid.: Road from Villa Tunari to Cochabamba, between Río Huayruruni and Río Huañuska, 17° 01' 02" S, 65° 33' 23" W, 433 m, 23 Oct. 2006, *Peters 06.0109* [SEL!; BGHD!]; ibid.: Road from Villa Tunari to Cochabamba, between Río Huayruruni and Río Huañuska, 17° 01' 02" S, 65° 33' 23" W, 433 m, 23 Oct. 2006, *Peters 06.0110* [FR!]; ibid.: Road from Villa Tunari to Cochabamba, between Río Huayruruni and Río Huañuska, 17° 01' 02" S, 65° 33' 23" W, 433 m, 23 Oct. 2006, *Peters 06.0111* [LPB!; BGHD!]; ibid.: Between Villa Tunari and Cristal Mayu, 17° 04' S, 65° 39' W, 580 m, 17 Oct. 2002, *Vásquez 4623* [FR!, VASQ!; FAN!]; ibid.: 128 km from Cochabamba to Villa Tunari, (17° 06' 00" S, 65° 40' 48" W), 630 m, 11 Aug. 1997, *Vásquez 2794* [VASQ!]; ibid.: Río Espiritu Santo, 640 m, 2 Sept. 1999, *Vásquez 3318 a* [FR!, LPB!, SEL!; FAN!]; ibid.: Between Río Huayruruni and Correo Huañuska, 17° 00' S, 65° 19' W, 440 m, 2 Sept. 1999, *Vásquez 3322* [FR!, LPB!; FAN!].

Distribution and ecology

Range size: Small.

Countries: BOLIVIA. Dpto. La Paz, Beni, Cochabamba.

Ecoregions: Sub-Andean-Amazon-Forests, Yungas.

Life style & habitats: Humid, evergreen lowland and montane forests. Terrestrial or saxicolous on steep slopes, in understory, along rocky, shaded roadsides and riverbanks. Sub-Andean-Amazon-Forests with *Pentaplaris davidsmithii* (Bombacaceae), Annonaceae, Lauraceae and Moraceae (*Fuentes 5366* [LPB]). Degraded Sub-Andean-Amazon-Forests with *Juglans boliviana* (Juglandaceae) (*Fuentes 9196* [LPB]). Dry slopes within Sub-Andean-Amazon-Forests with *Anadenanthera colubrina* (Fabaceae) (*Fuentes 6374* [LPB]).

Altitude: 400–1500 m.

Taxonomic delimitation and systematic relationships

Fosterella villosula is very similar to *F. christophii* and *F. micrantha* which all three are closely related but clearly distinct according to all available molecular data. *Fosterella villosula* varies in the densely villous inflorescence including floral bracts and sepals and longer floral bracts, sepals and petals. The three species also differ concerning their ecology, as *F. villosula* occurs in the very humid montane rain forests of the Chapare region, whereas *F. christophii* is restricted to more dry forests at the "Andean knee" close to the city of Santa Cruz and *F. micrantha* is to be found in Central America only.

One specimen has been collected in the "Área Natural de Manejo Integrado Madidi", which astonishingly has yellow petals (Bolivia. Dpto. La Paz: Prov. Franz Tamayo, Unapa, 21 km N of Apolo, 14° 32' 26" S, 68° 29' 46" W, 1022 m, 1 Sept. 2004, *Fuentes & Aldana 6374* [LPB!]). This seems to be quite unusual, but since all other character-states are consistent with *Fosterella villosula*, no further material and no molecular data is available for this specimen so far, it is not described as a new taxon.

Fig. 116: Lectotype of *Fosterella villosula* (Harms) L.B. Sm. [B].

Fig. 117: *F. villosula (Peters 06.0108)* in the natural habitat.

Fig. 118: Flowers of *F. villosula (Peters 06.0105)*.

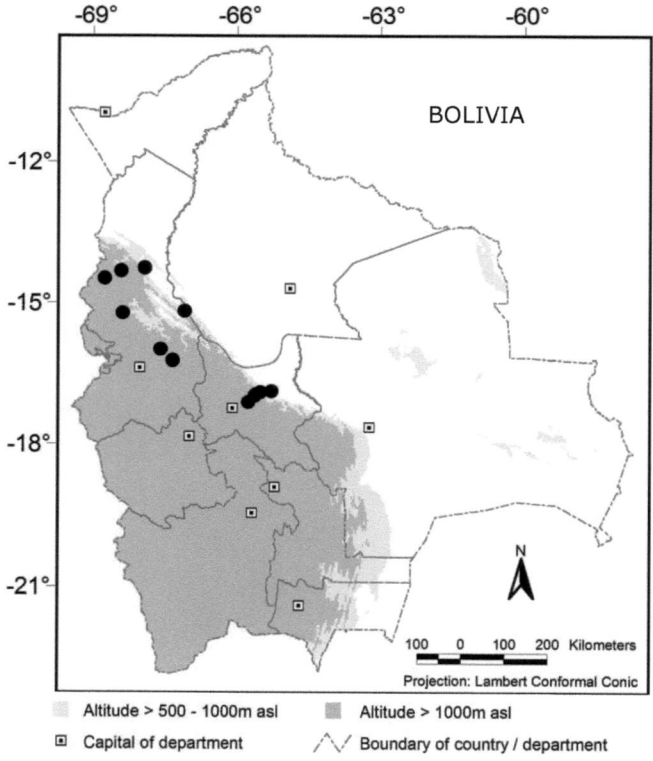

Fig. 119: Distribution of *Fosterella villosula*.

Fosterella weberbaueri (Mez) L.B. Sm., Phytologia 7: 172. 1960. Fig. 120–123.

Basionym:	*Lindmania weberbaueri* Mez, Repert. Spec. Nov. Regni Veg. 12: 417. 1913. TYPE: Perú. Dpto. Ayacucho: Prov. Huanta, next to Río Pieni (tributary stream of Río Apurímac), terrestrial on rocks at margin of rain forest, 800–1000 m, 2 June 1910, *Weberbauer 5635* [Holotype: B!].
=	*Fosterella schidosperma* var. *vestita* L.B. Sm. & Read, Phytologia 52 (1): 49. 1982. TYPE: Perú. Dpto. Puno: San Gabon to Ollachea, 1000–2000 m, 17–24 July 1978, *Dillon, Aronson, Herra & Berry 1253* [Holotype: US!; Isotype F!, MO!].

Comment on type

The correct name of the type locality of *Fosterella schidosperma* var. *vestita* L.B. Sm. & Read is "San Gabán" and not "San Gabon" as given in the original description. San Gabán is located in the Province of Carabaya in the Department of Puno at approx. 13° 20' S, 70° 20' W, 1000 m.

Etymology

This species is dedicated to August Weberbauer (1871–1948), collector of the type specimen, German botanist and pioneer of the investigation of the Peruvian flora.

Description

Plant subcaulescent, up to 90 cm high. **Leaves** few, up to 8–12, forming an open, flat rosette. **Sheaths** to 25 mm wide, entire, whitish, glabrous. **Blades** lanceolate, acuminate to acute, narrowed towards the base, 20–35 cm long and 2–5 cm wide, thin, entire, adaxially glabrous, abaxially scattered lepidote by peltate trichomes with dentate margin, glabrescent. **Peduncle** 20–60 cm long, to 2–4 mm in diameter, green, glabrous. **Peduncle bracts** 1–2 cm long, equalling the internodes, entire, abaxially glabrous. **Inflorescence** compound racemose, 25–30 cm long and 8–12 cm wide, axes green, slightly villous, glabrescent. **Primary bracts** 5–10 mm long, shorter than the sterile base of the branches, entire, abaxially glabrous. **Branches** 6–20, inclined, arcuate, 6–8 cm long, bearing up to 25 flowers. **Floral bracts** 1–2 mm long, shorter than or equalling the pedicels, entire, whitish, abaxially glabrous. **Flowers** spreading, suberect, 2–5 mm apart. **Pedicels** to 2 mm long. **Sepals** 2 mm long, white, glabrous. **Petals** 4–5 mm long, white, recoiled during anthesis but straight afterwards. **Filaments** 4 mm long. **Anthers** 1.5 mm long. **Style** 3–4 mm long. **Stigmatic complex** simple-erect. **Capsule** narrowly ovoid, 3–4 mm long, 2 mm wide. **Seeds** filiform, bicaudate, 2–3 mm long.

Specimens seen

PERÚ: **Prov. Huánuco:** 98 km from Huánuco towards Tingo Maria, 09°32' S, 75°57' W, 910 m, 3 June 1958, *Rahn 309* [MO!]; ibid.: Maquizapa, road to Monzón, 800 m, 23 Feb. 1966, *Schunke 1118* [NY!]; ibid.: Along the road from Huánuco to Tingo Maria, km 479, 09°34' S, 76°03' W, 1200 m, 1 June 1998, *Croat & Sizemore 81579* [MO!]. Dpto. Leoncio Prado: Rupa Rupa, Calpar Bella; next to Río Monzón and Cueva de los Hauriños, 800 m, 2 July 1976, *Schunke 9476* [F!, MO!, NY! (st.)]; ibid.: Rupa Rupa, W of Tingo Maria, hills facing the airport, (09° 08' S, 75° 57' W), 750 m, 1 Aug. 1978, *Schunke 10427* [MO!, SEL (phot!)]; ibid.: 26 km from Tingo Maria to Huánucu, between Las Palmas and Cayumba, (09° 30' S, 75° 54' W), 750 m, 1 June 1960, *Moore 8549* [US!]; ibid.: Rupa Rupa, Tingo Maria, road to Castillo, (09° 30' S, 75° 54' W), 800 m, 5 July 1978, *Plowman & Ramirez 7574* [F!, SEL (phot!), US!, USM!; MSBG!]. **Prov. Pasco:** Dpto. Oxapampa: Between Chatarra and Cacazu, 10° 32' S, 75° 04' W, 890 m, 10 July 2003, *Van der Werff 18212* [SEL!]. **Prov. Junin:** Dpto. Chanchmayo: Vitoc, 13 km from San Ramón, 11° 12' S, 75° 24' W, Aug. 1944, *Soukup 2446* [US!]. **Prov. Cuzco:** Dpto. La Convención: Trail along Río Mapituriani, a tributary of Río Apurimac, lowland rainforest opposite Hacienda Luisiana, 1000 m, (12° 48' S, 73° 36' W), 14 Sept. 1976, *Wasshausen & Encarnación 655* [US!]; ibid.: Echarate, Belenpata, 12° 49' S, 72° 35' W, 1750 m, 21 Aug. 2003, *Suclli & Farfan 1214* [SEL!]; ibid.: Vilcabamba, Chihuankiri, Esmeralda, 12° 48' 59" S, 73° 08' 58" W, 1117 m, 18 July 2204, *Calatayud 2492* [MO!, SEL! (st.)]; ibid.: Tinkuri, 800 m, 24 July 1948, *Bues s.n.* [CUZ!]; Vilcabamba, Chiwanquiri, 12° 45' 16" S, 73° 09' 00" W, 881 m, 18 July 2004, *Galiano 6587* [SEL!]. **Prov. Puno:** Dpto. Carabaya: San Gabán, (1000 m, 13° 20' S, 70° 20' W), July 1854, *Lechler 2382,* [G! (G 00074338) and K! (K 000321537)]; ibid. Below San Gabán, along Río San Gabán, 13° 24' S, 70° 18' W, 500–1000 m, 17–24 July 1978, *Dillon 1201* [F!, US!].
BOLIVIA: **Dpto. La Paz:** Prov. Franz Tamayo: Parque Nacional Madidi, between Carjata the old cemetery of Mojos, 14° 33' 48" S, 68° 52' 07" W, 1500 m, 1 July 2005, *Fuentes 9168 a* [LPB!]. Prov. Bautista Saavedra: Pauji-Yuyo, between Apolo and Charazani, 15° 02' S, 68° 29' W, 1150 m, 10 June 1997, *Kessler 10000* [LPB! (st.)]; ibid.: Area Natural de Manejo Integrado Apolobamba, Quita Calzon, 15° 09' 04" S, 68° 30' 44" W, 788 m, 10 Sept. 2004, *Cayola 1304* [LPB!, MO!]. Prov. Larecaja: Next to Río Tipuani, below Chusi, 15° 37' S, 68° 15' W, 1200 m, 14 Jan. 2001, *Müller 217/2* [SEL!, LPB!; FAN!]; ibid.: Between Guanay and Mapiri, 11 km N of La Aguada, 15° 28' 18" S, 67° 58' 18" N, 3 Aug. 2003, *Schinini 36437* [LPB!]. Prov. Caranavi: 34 km from Caranavi towards Sapecho, 15° 41' S, 67° 29' W, 1480 m, 26 Oct. 2002, *Vásquez 4656* [FAN!]. Prov. Nor Yungas: 15 km from Caranavi to Puerto Linares, next to Río Yara, 15° 42' S, 67° 29' W, 850 m, 15 Sept. 1981, *Beck 4810* [LPB!, SEL!, US!]; ibid.: Cotapata National Park, trails around Estación Biológica de Tunquini, 16° 11' S, 67° 52' W, 1550 m, 14 July 2000, *Krömer & Acebey 1230* [FR!, LPB!, SEL!]; ibid.: 4 km from Caranavi to Sapecho, 15° 40' 31" S, 67° 29' 17" W, 1506 m, 26 Oct. 2002, *Rex & Schulte 261002-7* [FR!]; ibid.: 4 km from Caranavi to Sapecho, 15° 40' 31" S, 67° 29' 17" W, 1506 m, 26 Oct. 2002, *Rex & Schulte 261002-8* [FR! (st.)]; ibid.: 4 km from Caranavi to Sapecho, 15° 40' 31" S, 67° 29' 17" W, 1506 m, 26 Oct. 2006, *Rex & Schulte 261002-9* [FR!]; ibid.: Valley of Río Unduavi, opposite to Santa Rosa, next to a waterfall, 1530 m, 5 Sept. 1987, *Vargas & Seidel 437* [LPB!]; ibid.: 15 km from Carasco to Sapecho, 15° 41' 19"S, 67° 29' 44"W, 1454 m, 26 Oct. 2002, *Rex & Schulte 261002-17* [SEL!]; ibid.: 15 km from Carasco to Sapecho, 15° 41' 19" S, 67° 29' 44" W, 1454 m, 26 Oct. 2002, *Rex & Schulte 261002-18* [FR! FAN!]; ibid.: 15 km from Carasco to Sapecho, 15° 41' 19" S, 67° 29' 44" W, 1454 m, 26 Oct. 2002, *Rex & Schulte 261002-19* [SEL!]. Prov. Sud Yungas: San Bartolome [near Callizaya], basin of Río Boopi, (16° 12' S, 67° 10' W), 750–900 m, July 1939, *Krukoff 10482* [F!, K!, LPB!, LIL!, MO!, NY!, U!, US!]. **Dpto. Cochabamba:** Prov. Chapare: Between San Rafael and El Palmar, 17° 05' S, 65° 29' W, 590 m, 8 Feb. 2000, *Vásquez 3570* [FR!, LPB!, SEL!, USZ!, VASQ!, WU!]; ibid.: Parque Nacional Carrasco, 17° 10' S, 65° 35' W, 1200 m, *Krömer 7286* [FR!, LPB!; BGHD!].

Distribution and ecology

Range size:	Medium.
Countries:	PERÚ: Dptc. Huánuco, Pasco, Junín, Cuzco, Ayacucho, Puno.
	BOLIVIA: Dpto. La Paz, Cochabamba.
Ecoregions:	Pre- and Sub-Andean-Amazon-Forests, Yungas.
Life style & habitats:	More ore less open, humid, evergreen lowland and montane forests. Terrestrial or saxicolous in shady understory, on steep, rocky slopes, on limestone hills, along roadsides or riverbanks on fallow fields. Degraded Sub-Andean-Amazon-Forests with *Juglans boliviana* (Juglandaceae) (*Fuentes 9168* a [LPB]). Evergreen forest with *Attalea* sp. and *Iriartea* sp. (Arecaceae) (*Kessler 10000* [LPB]).
Altitude:	600–1550 m.

Taxonomic delimitation and systematic relationships

IBISCH et al. (2002) already reinstated this taxon, which temporarily was combined to synonymy under *Fosterella schidosperma* (Baker) L.B. Sm. (SMITH 1963).

Actually, the two taxa *F. schidosperma* and *F. weberbaueri* are astonishingly similar and the taxonomic confusion has already begun with the heterogeneous type material of *Chlorophytum schidospermum* Baker (*Lechler 2382*), as pointed out above in the paragraph on *Fosterella schidosperma* (Baker) L.B. Sm. From this also follows that *Cottendorfia rushyi* Baker has to be combined to synonymy under *Fosterella schidosperma* (Baker) L.B. Sm.

Apart from the clear-cut floral character, that IBISCH et al. (2002) mentioned to delimit the two taxa (petals recoiled like watchsprings during anthesis and afterwards in *F. schidosperma* vs. only more or less recoiled petals that become straight after anthesis in *F. weberbaueri*), there are other distinctive features as well: the leaf blades of *F. weberbaueri* are rather short and wide, whereas in *F. schidosperma* they tend to be long and narrow. The inflorescence of *F. weberbaueri* is smaller, less ramified and slightly villous when very young, bearing spreading and suberect flowers.

As already explicated above, *Fosterella schidosperma* (Baker) L.B. Sm. var. *vestita* L.B. Sm. & Read is reduced to synonymy under *Fosterella weberbaueri* (Mez) L.B. Sm. here.

According to molecular data (REX et al. 2009), *F. weberbaueri* is closely related to *F. batistana* from the Brazilian Amazon. The two species obviously are morphologically related to each other, both characterised by few, rather broad, more or less glabrescent leaf blades forming a flat rosette, and more or less recoiled petals becoming straight postanthesis.

Fig. 120: Holotype of *Fosterella weberbaueri* (Mez) L.B. Sm. [B].

Fig. 121: *F. weberbaueri (Krömer 7286).*
Photo: T. Stolten.

Fig. 122: Flowers of *F. weberbaueri* *(Vásquez 4656).*

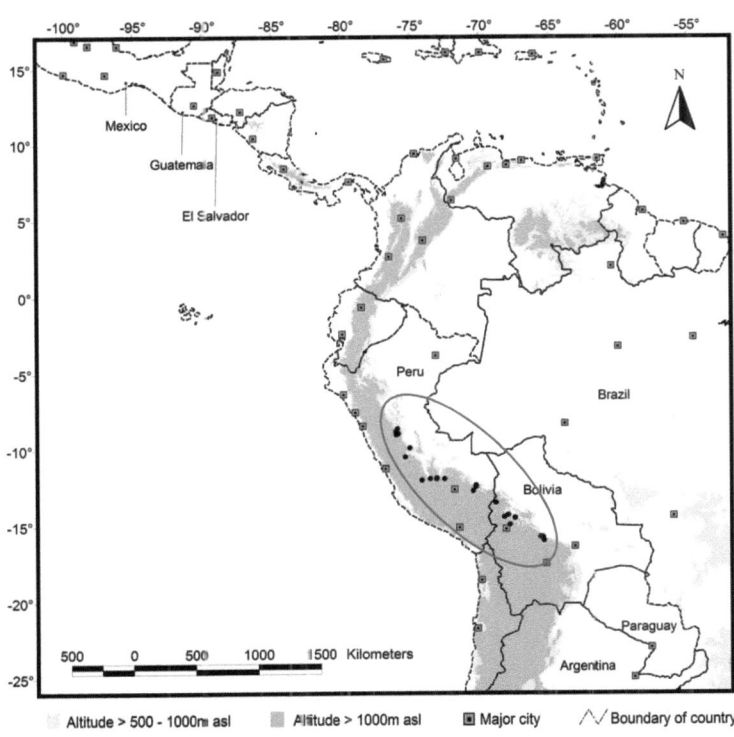

Fig. 123: Distribution of *Fosterella weberbaueri.*

Fosterella weddelliana (Brongn. ex Baker) L.B. Sm., Phytologia 7: 172. 1960. Fig. 124–128.

Basionym:	*Cottendorfia weddelliana* Brongn. ex Baker, Handbook of the Bromeliaceae: 129. 1889. TYPE: Bolivia. Yungas, 1200–2500 m, Dec. 1846, *Weddell 4233* [Lectotype: P (phot.!); Isotype: B!].
≡	*Lindmania weddelliana* (Brongn. ex Baker) Mez in C.DC., Monogr. phan. 9: 538. 1896.
=	*Fosterella nowickii* Ibisch, R. Vásquez & E. Gross, Selbyana 23 (2): 210. 2002. TYPE: Bolivia. Dpto. La Paz: Prov. Sud Yungas, nearby Irupana, 16° 26' S, 67° 28' W, 1200 m, 20 Feb. 2000 (cultivated specimen flowered in Oct. 2001), *C. Nowicki & R. Müller 2076* [Holotype: LPB!, USZ!]. Bolivia. Dpto. La Paz: Prov. Sud Yungas, 10 km from Irupana to Chulumani, 16° 26' S, 67° 28' W, 1400 m, 12 Feb. 2000, *R. Vásquez, G. Gerlach & L.R. Moreno 3636 a2* [Paratype: LPB!, SEL!].

Comment on type

BRONGNIART ex BAKER (1889) cited two specimens when he described *Cottendorfia weddelliana; Weddell 4233* and *Rusby 2541*, but did not choose a holotype. *Rusby 2541*, later was chosen as type for *Cottendorfia rusbyi* by BAKER (1902). Hence, the specimen *Weddell 4233* [P (phot.!)] had been designated as lectotype of *Fosterella weddelliana* (PETERS et al. 2008 b), as the isotype [B!] comprises only a fragment of a leaf.

A Clonotype of *Fosterella nowickii* Ibisch, R. Vásquez & E. Gross is cultivated in the living collection of the Fundación Amigos de la Naturaleza, Santa Cruz (Bolivia).

Etymology

This species is dedicated to Hugh Algernon Weddell (1819–1877) a British-French botanist specialised in South American flora and collector of the type specimen.

Description

Plant caulescent, up to 150 cm high. **Leaves** few 8–15, forming an open, arched rosette. **Sheaths** 50–70 mm wide, serrate, whitish, glabrous. **Blades** narrowly lanceolate, acuminate to acute, 15–50 cm long and 2–3.5 cm wide, succulent and serrate towards the base, adaxially lepidote towards the base, abaxially covered by a thick layer of interwoven, peltate trichomes with fimbriate margin. **Peduncle** 30–80 cm long, to 5 mm in diameter, green to reddish, glabrous, glaucous. **Peduncle bracts** 5–8 cm long, longer than the internodes, serrate, abaxially interwoven lepidote. **Inflorescence** paniculate with branches up to 2nd order, 20–40 cm long and 10–20 cm wide, axes green, glabrous, glaucous. **Primary bracts** 25–50 mm long, longer than or equalling the sterile base of the branches, entire to obscurely serrate, abaxially interwoven lepidote, glabrescent. **Branches** 8–25, ascending to inclined, arcuate, 10–15 cm long, bearing up to 30 flowers. **Secondary branches** to 5 cm long, bearing 5–8 flowers. **Floral**

bracts 2–3 mm long, shorter than the pedicels, entire, abaxially glabrous. **Flowers** secund, pendulous, 2–10 mm apart. **Pedicels** 2–4 mm long. **Sepals** 2–3 mm long, green to reddish, glabrous. **Petals** 4–5 mm long, white to greenish or reddish, recoiled like watchsprings during anthesis and afterwards. **Filaments** 3 mm long. **Anthers** 2–3 mm long. **Style** 2–4 mm long. **Stigmatic complex** simple-erect. **Capsule** narrowly ovoid, 4–5 mm long, 2–3 mm wide. **Seeds** filiform, bicaudate, 2–3 mm long.

Specimens seen

BOLIVIA: **Dpto. La Paz:** Prov. Franz Tamayo: Parque Nacional Madidi, between Puca Suchu and Virgen del Rosario, 14° 34' 40" S, 68° 41' 24" W, 910 m, 9 Nov. 2003. *Fuentes 5835* [LPB! (st.)]. Prov. Nor Yungas: 28 km on the road from Cotapata to Chulumani, valley of Río Unduavi 16° 21' 34" S, 67° 46' 18" W, 1890 m, 20 Dec. 2002, *Forzza 2370* [LPB! (fr.), RB! (fr.)]; ibid.: 42 km on the road from Coroico to Caranavi, between Santa Barbara and San Pedro, 16° 01' 13" S, 67° 37' 20" W, 864 m, 20 Oct. 2006, *Peters 06.0092* [FR!]; ibid.: 42 km on the road from Coroico to Caranavi, between Santa Barbara and San Pedro, 16° 01' 13" S, 67° 37' 20" W, 864 m 20 Oct. 2006, *Peters 06.0093* [SEL!]; ibid.: 42 km on the road from Coroico to Caranavi, between Santa Barbara and San Pedro, 16° 01' 13" S, 67° 37' 20" W, 864 m, 20 Oct. 2006, *Peters 06.0094* [LPB!]; ibid.: About 9 km N of Yalosa, on the road to Caranavi, 16° 12' 06" S, 67° 44' 08" W, 1250 m, 9 Mar. 1984, *Varadarajan 1289* [GH! (fr.)]. Prov. Sud Yungas: Nearby Irupana, 16° 25' 48" S, 67° 28' 12" W, 1192 m, 25 Feb. 2000, *Nowicki & Müller 2075* [FR!, VASQ!]; ibid.: Road from Unduavi to Puente Villa, along Río Unduavi, 16° 21' 51" S, 64° 23' 47" W, 1804 m, 18 Oct. 2006, *Peters 06.0064* [LPB!; BGHD!]; ibid.: Road from Unduavi to Puente Villa, along Río Unduavi, 16° 21' 51" S, 64° 23' 47' W, 1804 m, 18 Oct. 2006, *Peters 06.0065* [LPB!]; ibid.: Road from Unduavi to Puente Villa, along Río Unduavi, 16° 21' 51" S, 64° 23' 47" W, 1804 m, 18 Oct. 2006, *Peters 06.0066* [BGHD!]; ibid.: Road from Unduavi to Puente Villa, along Río Unduavi, 16° 21' 51" S, 64° 23' 47" W, 1804 m, 18 Oct. 2006, *Peters 06.0067* [SEL!]; ibid.: Road from Unduavi to Puente Villa, along Río Unduavi, 16° 21' 51" S, 64° 23' 47" W, 1804 m, 18 Oct. 2006, *Peters 06.0068* [LPB!]; ibid.: Road from Unduavi to Puente Villa, along Río Unduavi, 16° 24' 08" S, 67° 38' 46" W, 1313 m, 18 Oct. 2006, *Peters 06.0069* [LPB!]; ibid.: 19 km on the road from Chulumani to Choquechaca, along Río Unduavi, behind Tajma, 16° 21' 26" S, 67° 27' 51" W, 1463 m, 19 Oct. 2006, *Peters 06.0079* [LPB! (st.)]; ibid.: 19 km on the road from Chulumani to Choquechaca, along Río Unduavi, behind Tajma, 16° 21' 26" S, 67° 27' 51" W, 1463 m, 19 Oct. 2006, *Peters 06. 0081* [SEL!]; ibid.: 37 km from Chulumani towards Choquechaca, along the Unduavi, behind Tajma, 16° 17' 27" S, 67° 21' 25" W, 930 m, 19 Oct. 2006, *Peters 06.0083* [FR! (st.)]; ibid.: 78 km from Chulumani towards La Asunta, along Río Unduavi, behind Choquechaca, 16° 10' 17" S, 67° 10' 26" W, 756 m, 19 Oct. 2006, *Peters 06.0084* [FR!]; ibid.: 14 km from Chulumani towards Puente Villa, 16° 22' 36" S, 67° 35' 41" W, 1462 m, 20 Oct. 2006, *Peters 06.0087* [LPB!]; ibid.: 14 km from Chulumani towards Puente Villa, 16° 22' 36" S, 67° 35' 41" W, 1462 m, 20 Oct. 2006, *Peters 06.0088* [SEL!]; ibid.: 9 km from Chamaca towards Chulumani, Río La Paz, 16° 17' S, 67° 17' W, 900 m, 3 Oct. 1995, *Kessler 5727* [SEL!]; ibid.: 10 km from Irupana towards Chulumani, 16° 26' S, 67° 28' W, 1400 m, 12 Feb. 2000, *Vásquez 3636* [FR!, LPB!, SEL!]; ibid.: 10 km from Irupana towards Chulumani, 16° 26' S, 67° 28' W, 1400 m, 12 Feb. 2000, *Vásquez 3636 d* [LPB!; FAN!]; ibid.: Between Unduavi and Yanacachi, 3 km before the fork, 16° 21' S, 67° 48' W, 1820 m, 9 Sept. 1989, *Beck 16862* [LPB!, SEL!]; ibid.: 9 km E of Puente Villa on the road to Chuluman, 16° 23' S, 67° 36' W, 1400 m, 29 Sept. 1985, *Solomon & Nee 14316* [LPB!, MC!, SEL! (fr.)]; ibid.: Cieneguillas, nearby La Plazuela, valley of Río La Paz, 16° 35' S, 67° 26' W, 1300 m, 14 Dec. 1997, *Beck 23136* [LPB!]; ibid.: Below Villa Barrientos, valley of Río Tamanpaya, above Totorapampa, 1100 m, 3 Jan. 1990, *Beck 17338* [LPB!, SEL! (fr.)]; ibid.: 26 km from Chulumani towards Asunta, behind Tajma, 1300 m, 27 June 1985, *Beck 12078* [LPB!]; ibid.: 30 km from Chulumani towards Asunta and Colquechaca, union of Río Solacama and Río Tamanpayo, 1250 m, 29 June 1985, *Beck 12168* [LPB! (st.)]; ibid.: Between Chulumani and La Asunta, Chamaca, next to Río Boopi, 16° 14' S, 67° 14' W, 900 m, 21 Sept. 1999, *Beck 25104* [FR! (st.), LPB!, SEL!]; ibid.: Road from Miguillas to Plazuela, 16° 32' S, 67° 22' W, 1550 m, 14 Nov. 1998, *Beck PG98-50* [LPB!]; ibid.: La Plazuela, behind the bridge across Río La Paz, 16° 32' S, 67° 23' W, 1200 m, 15 Dec. 1998, *Beck 23232* [LPB! (fr.)]; ibid.: Between Asunta and Choquechalca, union of Río Solacama and Río Tamanpayo, 16° 06' S, 67° 12' W, 1300 m, 16 Dec. 1997, *Beck 23278* [LPB!]. Prov. Inquisivi: 16° 48' 00" S, 67° 12'

36" W, 1846 m, 24 Feb. 2000, *Nowicki & Müller 2034* [FR!; FAN!]; ibid.: 11 km from Inquisivi towards Circuata, 16° 53' S, 67° 08' W, 2170 m, 11 Feb. 2000, *Váquez 3612* [FR!, SEL!, VASQ!]; ibid.: Nearby Bolea, 16° 41' S, 67° 14' W, 2720 m, 11 Feb. 2000, *Vásquez 3627* [FR!; FAN!]; ibid.: Miguillas, between Irupana and Cajuata, next to Río La Paz, 16° 32' 24" S, 67° 22' 48" W, 1353 m, 24 Feb. 2000, *Nowicki 2060* [FR!, LPB!; FAN!].

Distribution and ecology

Range size: Small.

Country: BOLIVIA. Dpto. La Paz.

Ecoregions: Sub-Andean-Amazon-Forests, Yungas, Inter-Andean-Dry-Forests.

Life style & habitats: Humid, evergreen lowland and montane forests and dry, deciduous valleys. Terrestrial or saxicolous on steep, rocky slopes along roadsides. Semideciduous forest with *Anadenanthera colubrina* (Fabaceae) (*Kessler 5727* [SEL]). Within a forest dominated by *Schinopsis* sp. (Anacardiaceae), on rocky slope (*Beck 17338* [LPB]). Semideciduous forest on rocky slopes with *Baccharis* sp. (Asteraceae) and *Gynerium* sp. (Poaceae) (*Fuentes 5835* [LPB]).

Altitude: 750–2750 m.

Taxonomic delimitation and systematic relationships

Because of its very fragmentary type material and therefore misleading original description, *Fosterella weddelliana* was difficult to interpret. BAKER (1889) considered the plant to be acaulescent, a character that not has been questioned by later authors, neither by SMITH & DOWNS (1974), SMITH & READ (1992), nor by R.W. Read in his unpublished key from the 1990s. After some confusion about the identity of both *F. weddelliana* and *F. nowickii*, careful comparison of the leaves, inflorescences and flowers of the type material, it became clear that the two can not be separated. Therefore, *F. nowickii* was reduced to synonymy of *F. weddelliana* (PETERS et al. 2008 b), which is clearly caulescent with flowering plants reaching heights of up to 150 cm. It tends to form large mats covering steep slopes. According to molecular data (REX et al. 2009), *Fosterella weddelliana* is very close to *F. cotacajensis*, which morphologically is very much alike. But compared to *F. cotacajensis*, *F. weddelliana* has broader leaf blades, which are narrowed towards the base and shorter floral bracts and sepals. Molecular data also indicate a close relation as well as a clear separation of these two species, which are forming a well defined and supported group (REX et al. 2009).

F. cotacajensis differs from *F. rusbyi* – to which it is quite similar as well – in narrower leaf blades, longer floral bracts and sepals, serrate peduncle bracts and pronounced caulescence.

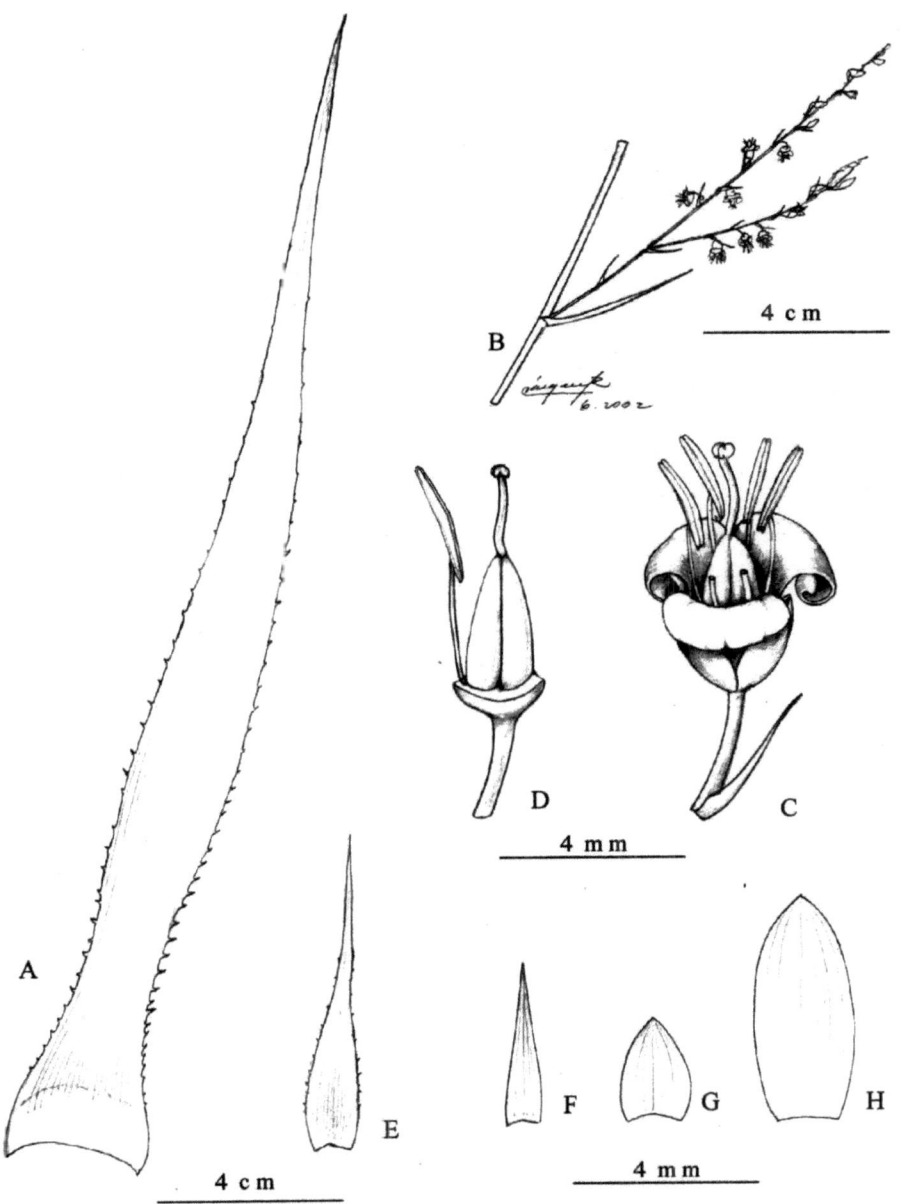

Fig. 124: *Fosterella weddelliana*. A Leaf. B Inflorescence branch and primary bract. C Flower. D Gynoecium and stamen. E Scape bract. F Floral bract. G Sepal. H Petal. Drawing of *Nowicki & Müller 2076* by R. Vásquez in IBISCH et al. (2002). © Selbyana.

Fig. 125: Lectotype of *Fosterella weddelliana* (Brongn. ex Baker) L.B. Sm. [P].

Fig. 126: *F. weddelliana (Peters 06.0079)* in the natural habitat.

Fig. 127: Flowers of *F. weddelliana (Peters 06.0087)*.

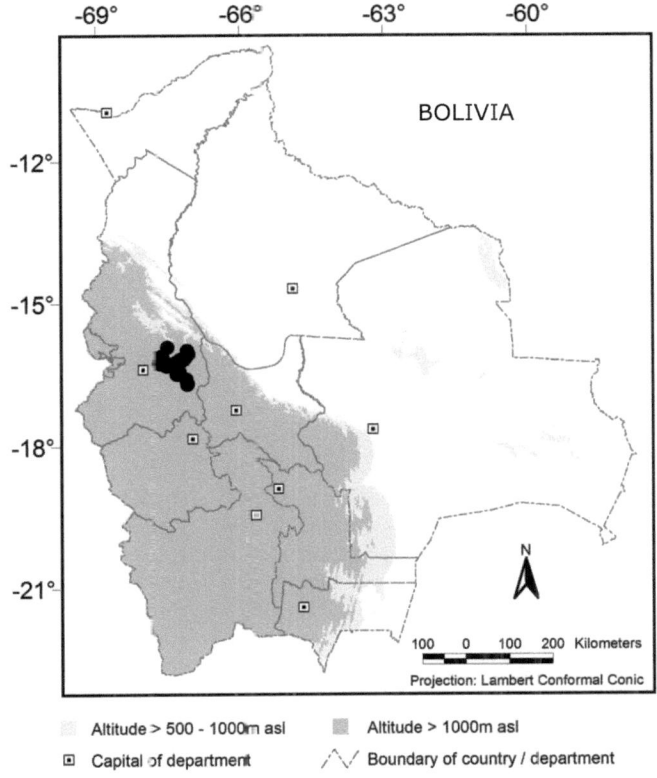

Fig. 128: Distribution of *Fosterella weddelliana*.

Fosterella windischii L.B. Sm. & Read, Bradea 6 (15): 137. 1992. Fig. 129–132.

> TYPE: Brazil. Estado Mato Grosso: Serra Ricardo Franco, crevices in woody slopes, near the Bolivian border, 450–500 m, 15° S, 60° W, 23 Sep. 1978, *P.G. Windisch 2044* [Holotype: US (phot!); Isotype: HB!].

Etymology

This species is dedicated to Paolo Günther Windisch (*1948), Brazilian botanist and collector of the type specimen.

Description

Plant acaulescent, up to 60 cm high. **Leaves** few, 6–12, forming an open, more or less upright rosette. **Sheaths** to 25 mm wide, entire, whitish, glabrous. **Blades** narrowly lanceolate, acuminate, narrowed towards the base, 30–50 cm long and 1.5–2.5 cm wide, succulent towards the base, entire or obscurely serrate towards the base, adaxially glabrous, abaxially covered by a thick layer of interwoven, peltate trichomes with fimbriate margin. **Peduncle** to 30 cm long, to 3 mm in diameter, green, glabrous. **Peduncle bracts** 2 cm long, longer than the internodes, entire, abaxially interwoven lepidote. **Inflorescence** compound racemose, 15–25 cm long and 6–10 cm wide, axes green, glabrous. **Primary bracts** 5 mm long, shorter than the sterile base of the branches, entire, abaxially interwoven lepidote, glabrescent. **Branches** 6–15, ascending, straight, 6–10 cm long, bearing up to 20 flowers. **Floral bracts** 1–2 mm long, shorter than the pedicels, entire, abaxially glabrous. **Flowers** secund, pendulous, 5 mm apart. **Pedicels** 2–3 mm long. **Sepals** 2–3 mm long, green, glabrous. **Petals** 4–5 mm long, white, recoiled like watchsprings during anthesis and afterwards. **Filaments** 3–4 mm long. **Anthers** 1.5–2 mm long. **Style** 2 mm long. **Stigmatic complex** simple-erect. **Capsule** ovoid, 4–5 mm long, 2–2.5 mm wide. **Seeds** filiform, bicaudate, 2–3 mm long.

Specimens seen

BOLIVIA: **Dpto. Santa Cruz:** Prov. Velasco: Parque Nacional Noel Kempff, Lago Caimán, 2 km uphill S of the camp, 13° 36' 54" S, 60° 54' 36" W, 300 m, 2 Oct. 1995, *Vargas 4001* [MO!]; ibid.: Parque Nacional Noel Kempff, Las Torres, 13°42'55" S, 61°41'43" W, 19 July 2003, *Ibisch 03.0016* [LPB!; BGHD!].
BRAZIL: **Estado Mato Grosso:** Without precise locality, 3 Sept. 1978, *Burlemax 67206* [HB!]; ibid.: Sierra Ricardo Franco, crevices in woody slopes, near the Bolivian border, 15° S, 60° W, 450–500 m, July 1957, *Windisch 1326* [HB!]; ibid.: Serra Ricardo Franco, 15° S, 60° W, 400–450 m, 13 Dec. 1977, *Windisch 1482* [HB! (fr.)]; ibid.: Vila Bela da Santíssima Trindade, Serra de Ricardo Franco, Cachoeira dos Namorados, riverbanks, 13 July 2007, *Krantz 212* [LEME 7144!]; ibid.: Vila Bela da Santíssima Trindade, Serra de Ricardo Franco, Cachoeira Seca, on vertical shaded wall, 13 July 2007, *Krantz 213* [LEME 7145!]. Município Bonito: Farm Baía das Garcas, Waterfall of Río Aquidaban, 550 m, 12 Nov. 2002, *Hatschbach 74128* [RB!].

Distribution and ecology

Range size:	Very small.
Countries:	BOLIVIA. Dpto. Santa Cruz. BRAZIL. Estado Mato Grosso.
Ecoregions:	Beni-and-Santa-Cruz-Amazon-Forests.
Life style & habitats:	Dry rock slopes of the Caparús table mountain (Precambrain rocks of the Brazilian shield) within humid, evergreen lowland forests.
Altitude:	300–550 m.

Taxonomic delimitation and systematic relationships

Morphologically, *Fosterella windischii* is quite similar to *F. kroemeri* but differs in shorter and narrower leaf blades whith more pronounced indumentum on the abaxial leaf surface, shorter peduncles and less ramified inflorescence.

Molecular results (REX et al. 2009) show that the two species are not closely related with each other. According to the molecular trees, *F. windischii* belongs to a group growing on Precambrian outcrops including *F. vasquezii* and *F. yuvinkae*, while *F. kroemeri* groups together with *F. caulescens*, *F. rexiae* and *F. albicans*. Accordingly, there are considerable ecological differences, as *F. kroemeri* occurs in the humid montane rain forests of Bolivia and *F. windischii* is confined to a small area of Precambrian rocks of the Brazilian shield.

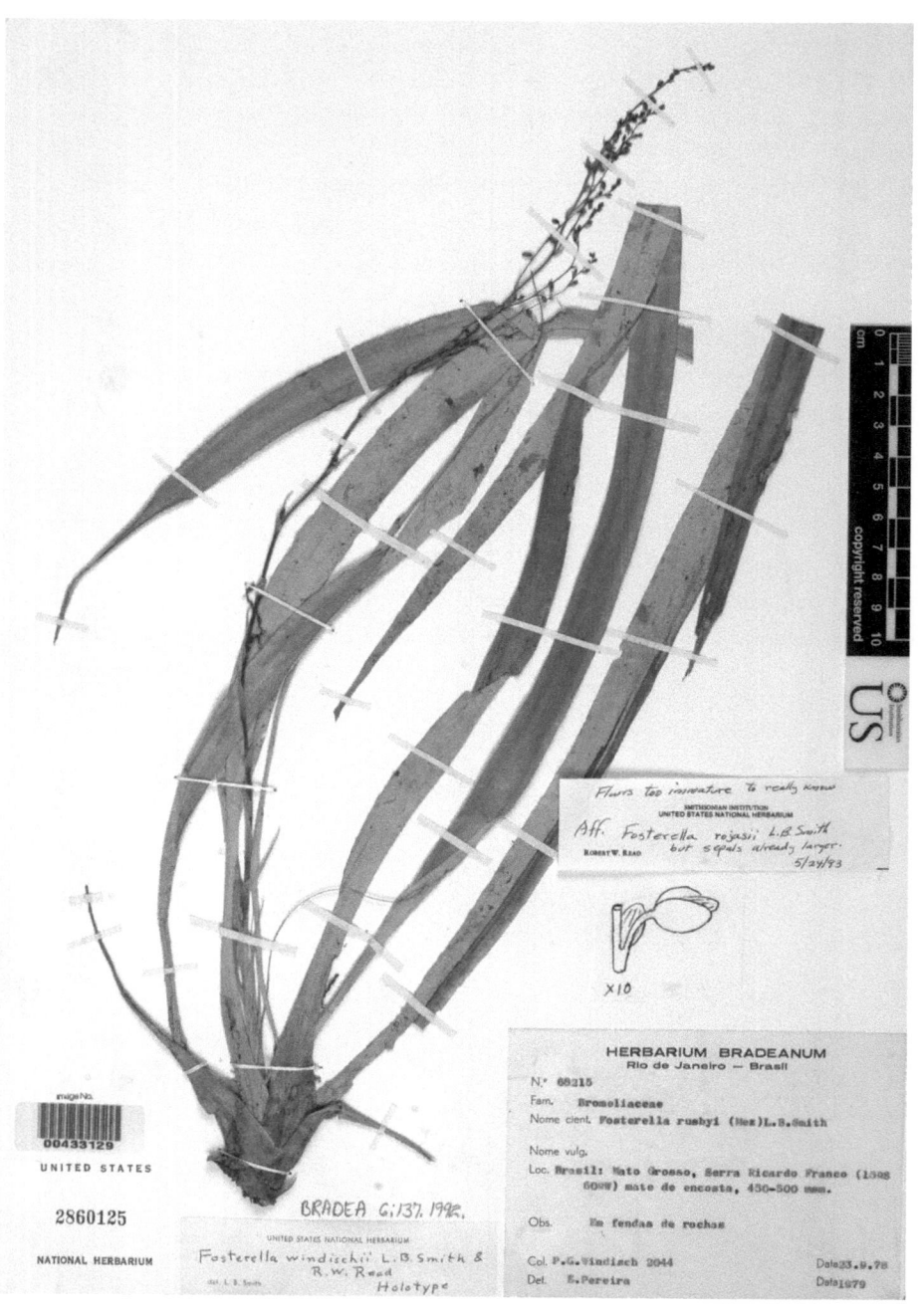

Fig. 129: Holotype of *Fosterella windischii* L.B. Sm. & Read [US]. © Smithsonian National Museum of Natural History.

Fig. 130: *F. windischii (Ibisch 03.0016)* in the natural habitat. Photo: P. Ibisch.

Fig. 131: Flowers of *F. windischii (Krantz 213)*.

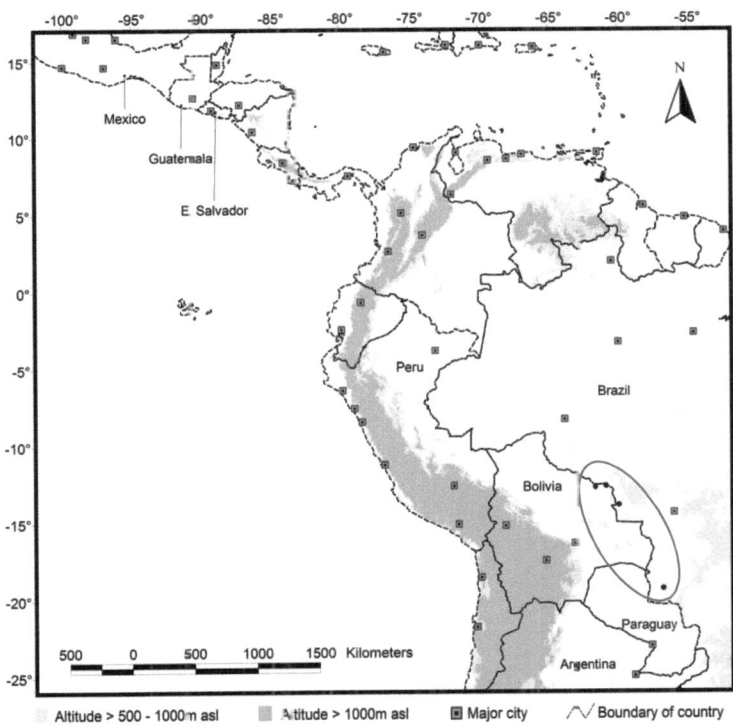

Fig. 132: Distribution of *Fosterella windischii*.

Fosterella yuvinkae Ibisch, R. Vásquez, E. Gross & S. Reichle, Selbyana 23 (2): 216. 2002.
Fig. 133–137. TYPE: Bolivia. Dpto. Santa Cruz: Prov. Chiquitos, Santiago de Chiquitos, 18° 20' 108" S, 59° 34' 301" W, Aug. 2001, *Reichle P-SR1* [Holotype: LPB!].

Comment on type

The type collection was made on steep slopes beside rocky rivulets in riverine forest of the Santiago mountain chain within the Chiquitano Dry Forest ecoregion, where plants either grew on rocks or directly on the ground (IBISCH et al. 2002).

Clonotypes are cultivated in the Botanical Gardens Heidelberg (Germany) and in the living collection of the Fundación Amigos de la Naturaleza, Santa Cruz (Bolivia).

Etymology

This species is dedicated to the Bolivian zoologist Yuvinka Gareca (*1978), in honour for her ecological research in the area of Santiago de Chiquitos.

Description

Plant acaulescent, up to 40 cm high. **Leaves** many, 15–20, forming a open, flat rosette. **Sheaths** 20–30 mm wide, entire, whitish, villous. **Blades** linear to narrowly oblanceolate, acuminate, narrowed towards the base, 20–35 cm long and 1.5–3 cm wide, succulent towards the base, entire, adaxially glabrous, abaxially densely tomentose by sessile, stellate trichomes. **Peduncle** 30–50 cm long, 3–4 mm in diameter, green, glabrous, glaucous. **Peduncle bracts** 2–4 cm long, longer than the internodes, entire, abaxially villous. **Inflorescence** paniculate with branches up to 2nd order, 15–30 cm long and 6–18 cm wide, axes green, glabrous, glaucous. **Primary bracts** 5–10 mm long, shorter than the sterile base of the branches, entire, abaxially villous, glabrescent. **Branches** to 15, inclined, arcuate, 8–20 cm long, bearing up to 20 flowers. **Secondary branches** 5–7 cm long, bearing 5–7 flowers. **Floral bracts** 2 mm long, shorter than or equalling the pedicels, entire, abaxially glabrous. **Flowers** secund, pendulous, 5–10 mm apart. **Pedicels** 2–4 mm long. **Sepals** 3–4 mm long, green, glabrous. **Petals** 7–9 mm long, very narrow, 1.5 mm wide, white, recurved during anthesis, straight afterwards. **Filaments** 4–6 mm long. **Anthers** 2 mm long. **Style** 5 mm long. **Stigmatic complex** conduplicate-spiral. **Capsule** ovoid, 5 mm long, 3 mm wide. **Seeds** filiform, bicaudate, 3–4 mm long.

Specimens seen

BOLIVIA: **Dpto. Santa Cruz:** <u>Prov. Chiquitos:</u> Santiago de Chiquitos, Serrania de Santiago, 18° 40' S, 59° 15' W, 800 m, 13 Nov. 1997, *Mamani & Jardim 1261* [MO! (fr.), SEL! (fr.), USZ! (fr.), WU! (fr.)]; ibid.: Santiago de Chiquitos, Serrania de Santiago, sector La Cruz, 18° 19' 35" S, 59° 33' 57" W, 700 m, 20 Nov. 2001, *Guillén 5671* [SEL!]; ibid.: Santiago de Chiquitos, Serrenia de San Miserato, (18° 24' S, 59° 36' W), 850 m, 13 July 2002, *Vásquez 4545* [SEL!]; ibid.: Between Santiago de Chiquitos and Sanjuanoma, 18° 19' S, 59° 37' W, 670 m, 11 July 2002, *Vásquez 4510*

[FR!, HEID!, LPB! SEL!; FAN!]; ibid.: Between Roboré and Santiago de Chiquitos, 416 m, Oct. 1934, *Cardenas 2994* [LIL!]; ibid.: Serrania de Santiago, 18° 20' 27" S, 59° 33' 57" W, 650 m, 27 Nov. 1999, *Guillén 4829* [MO!]; ibid.: Serrania de Santiago, 17 km from Santiago de Chiquitos towards Roboré, 18° 20' S, 59° 40' W, 480 m, Oct. 1995, *Amerhauser 5* [WU!; HBV]. Prov. Cordillera: Lagunillas, Cordillera de Cara, 19° 39' S, 63° 40' W, 1200 m, Aug. 1934, *Cardenas 2856 b* [F!].

Distribution and ecology

Range size: Very small.

Country: BOLIVIA. Dpto. Santa Cruz.

Ecoregion: Chiquitano-Dry-Forest.

Life style & habitats: Precambrian rock slopes of the Serranía de Santiago (Brazilian shield) within dry, deciduous lowland forests. Steep, rocky slopes, on sandy ground, along small ravines and riverbanks.

Altitude: 400–1200 m.

Taxonomic delimitation and systematic relationships

Fosterella yuvinkae is similar to *F. penduliflora* but differs in densely tomentose abaxial leaf surface. While *F. penduliflora* is widely distributed in the montane forests of Bolivia, *F. yuvinkae* occurs on Precambrian rock outcrops within the Chiquitano-Dry-Forests of Bolivia.

Furthermore *F. yuvinkae* resembles *F. hatschbachii*, but differs in bigger stature; more leaves, longer peduncle, more branched inflorescence, longer primary bracts, branches, floral bracts and petals. Moreover *F. yuvinkae* differs from all other *Fosterella* species by its very narrow petals.

According to molecular data, *F. yuvinkae*, *F. vasquezii* and *F. windischii* form a group of species from the Brazilian shield, which morphologically are clearly distinct from each other.

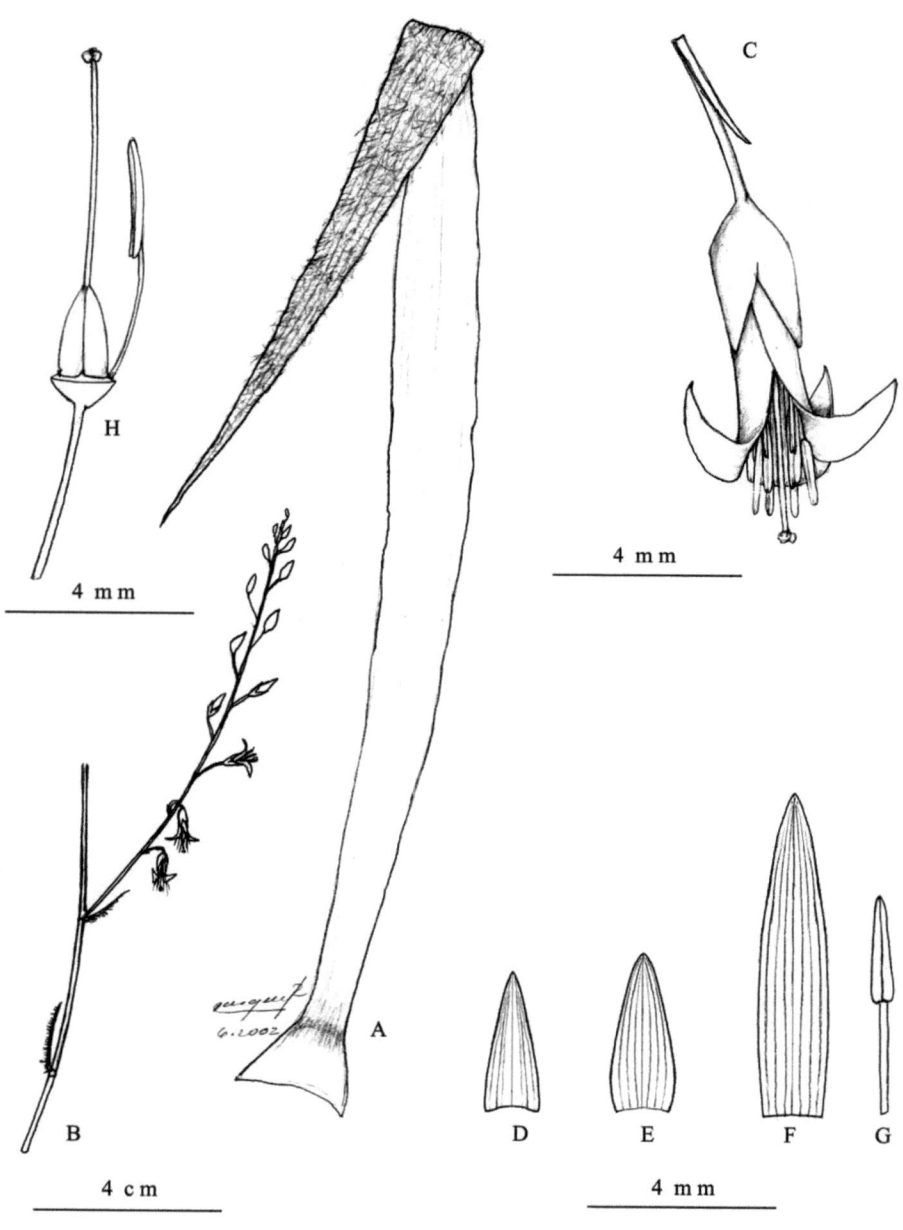

Fig. 133: *Fosterella yuvinkae*. A Leaf. B Inflorescence branch and primary bract. C Flower. D Floral bract. E Sepal. F Petal and stamen. G Gynoecium and stamen. Drawing of *Reichle P-SR1* by R. Vásquez in IBISCH et al. (2002). © Selbyana.

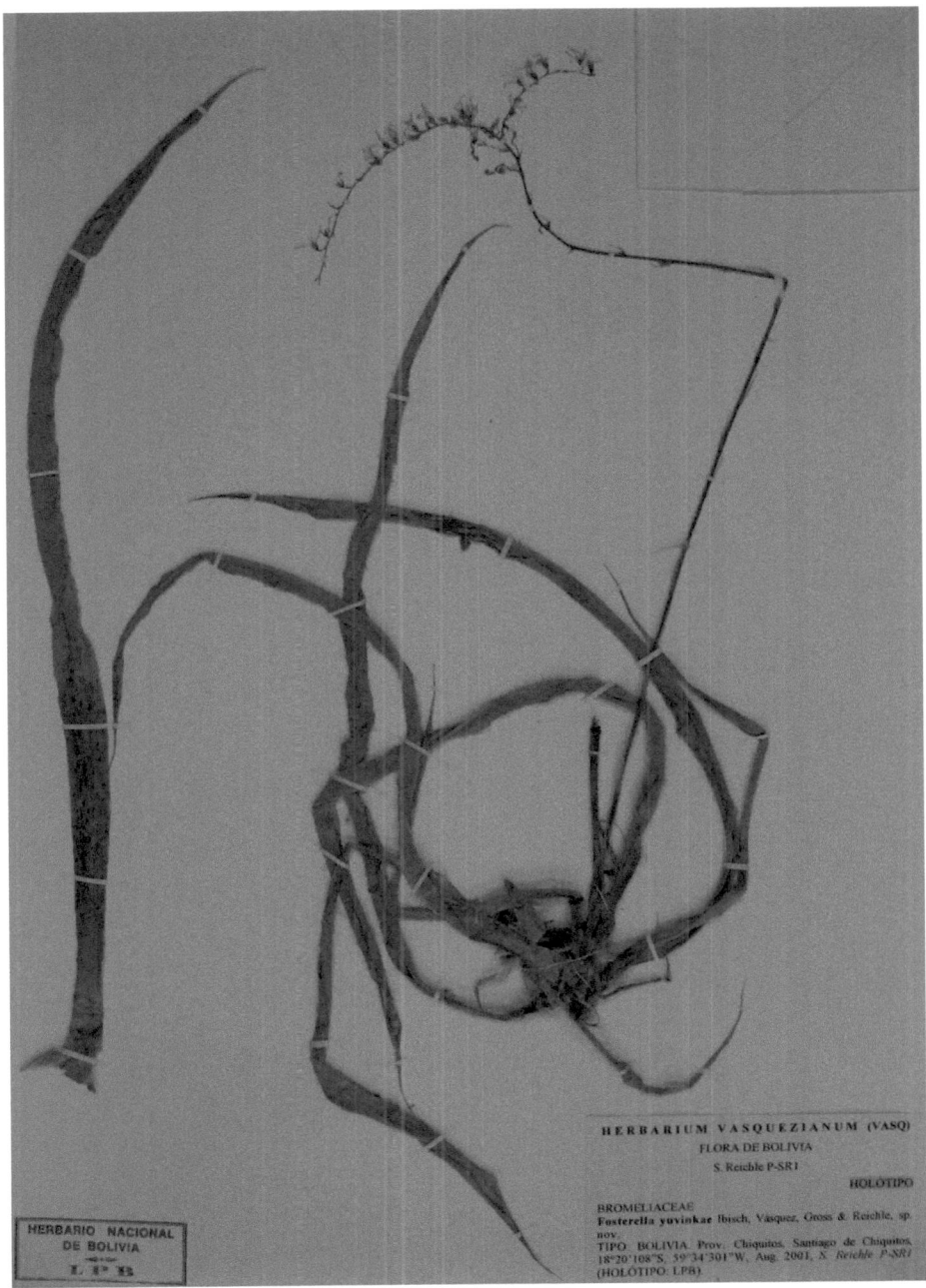

Fig. 134: Holotype of *Fosterella yuvinkae* Ibisch, R. Vásquez, E. Gross & S. Reichle [LPB].

Fig. 135: *F. yuvinkae (Reichle P-SR1).*

Fig. 136: Flowers of *F. yuvinkae (Reichle P-SR1).*

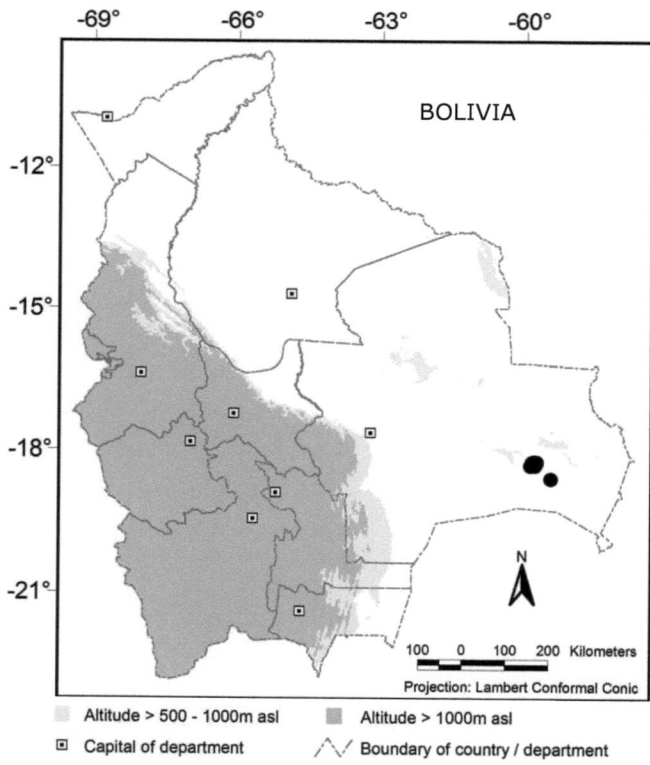

Fig. 137: Distribution of *Fosterella yuvinkae.*

***Fosterella* spp.** (indet., material too poor to determine)

BOLIVIA: **Dpto. La Paz:** Prov. Franz Tamayo: Parque Nacional Madidi, Río Hondo, Arroyo Negro, Serrania de Toregua, 14° 40' 21" S, 67° 50' 39" W, 650 m, 31 Mar. 2002, *Fuentes 4481* [LPB! (st.)]; ibid.: Parque Nacional Madidi, track from Apolo to Azariamas, Arroyo Pintata, 14° 27' 54" S, 68° 32' 56" W, 938 m, 3 Mar. 2003, *Canqui 253* [LPB! (st.), SEL! (st.)]. Prov. Larecaja: 60 km from Caranavi to Guanay, 550 m, *Besse s.n.* [SEL (phot.!); MSBG]. Prov. Nor Yungas: Above Yolosilla, 16° 12' S, 67° 44' W, 1250 m, 20 May 1995, *Kessler 4214* [SEL! (st.)]; ibid.: 16 km SW of Yalosa, on the road from Chuspipata to Coroico, 16° 15' S, 67° 46' W, 1850 m, 8 Mar. 1984, *Varadarajan 1282* [GH! (fr.), MO! (fr.)]; ibid.: 2 km from Choro towards San Pedro, next to Rio Alto Choro, 16° 01' 48" S, 67° 37' 18" W, 800 m, 26 Oct. 2002, *Rex & Schulte 261002-2* [FR! (st.)]. Prov. Sud Yungas: Irupana, 22 km towards the new bridge across Río La Paz, 1100 m, 18 May 1980, *Beck 2951* [LPB! (st.)]. Prov. Inquisivi: 15 km from Circuata to Chulumani, 1370 m, *Besse 1843* [SEL (phot.!); MSBG]. **Dpto. Beni:** Prov. Moxos: Parque Nacional Isoboro Secure, 3 km W of Oromono, 15° 51' S, 66° 22' W, 250 m, 19 May 1992, *Paz 97 A* [SEL! (st.)]. **Dpto. Chuquisaca:** Prov. Jaime Mendoza: ibid.: 98 km from Monteagudo towards Padilla, 19° 32' S, 64° 10' W, 1500 m, 2 July 1995, *Kessler 5031* [SEL! (st.)]. **Dpto. Santa Cruz:** Prov. Vallegrande: 17 km from Loma Larga to Masicurí, 18° 48' S, 63° 49' W, 1000 m, 24 May 1996, *Kessler 6048* [SEL! (st.), LPB! (st.)].
BRAZIL: **Estado Mato Grosso:** Município Corumbá: Morro do Urucum, 11 Nov. 1977, *H.C. de Lima* 132 [RB! (st.)].
WHITHOUT LOCALITY: *s.n.* [WU 8558! (st.); HBV B190/96 ex BGB].

3.4 Species ranges and diversity

The majority of *Fosterella* species have relatively small distribution ranges, and are also restricted to isolated and/or specific habitats like inter-Andean dry valleys or rock outcrops within in the lowlands. These species are *Fosterella aletroides*, *F. batistana*, *F. christophii*, *F. cotacajensis*, *F. elviragrossiae*, *F. floridensis*, *F. graminea*, *F. hatschbachii*, *F. heterophylla*, *F. kroemeri*, *F. nicoliana*, *F. rexiae*, *F. robertreadii*, *F. rojasii*, *F. vasquezii*, *F. windischii* and *F. yuvinkae*.

The following *Fosterella* species have a medium-sized distribution range: *Fosterella caulescens*, *F. chaparensis*, *F. gracilis*, *F. pearcei*, *F. petiolata*, *F. rusbyi*, *F. schidosperma*, *F. spectabilis*, *F. villosula*, *F. weberbaueri* and *F. weddelliana*. Most of these are native to the humid montane Yungas rain forests of northern Bolivia.

Only three *Fosterella* species are wide-spread and show large geographical ranges: *Fosterella albicans*, *F. micrantha* and *F. penduliflora*. *Fosterella micrantha* grows in more or less dry tropical to subtropical forests in Central America, from Mexico in the North to El Salvador in the South. Both *F. albicans* and *F. penduliflora* occur in subtropical as well as tropical forests (Yungas and Tucuman-Bolivian Forest) along the eastern slopes of the Andes from the North of Bolivia to Argentina in the South. Hence they belong to the few *Fosterella* species, which are distributed north and south of the Andean knee at the latitude of Santa Cruz. Furthermore, *F. penduliflora* has a disjunct distribution in the Chiquitano-Dry-Forests within the Bolivian lowlands, where it is delimited to azonal sites associated with the Brazilian shield (Precambrian rock outcrops).

Contrary to the common perception of *Fosterella* being a genus of rather dry habitats, about half of the recently described taxa prefer humid sites and many of the known species are true rain forest species. Thus, the cloud forest understory species *F. rusbyi* represents the genus' cool-humid limit of ecological preference. The cool-arid limit is marked by *F. cotacajensis*, which grows in high inter-Andean dry valleys. Most *Fosterella* species occur along the Eastern slopes of the Central Andes.

The highest species diversity is found in the ecoregion of the Peruvian-Bolivian Yungas, particularly in the La Paz department of Bolivia. A second diversity center is the Pre-Cambrian shield, where *Fosterella windischii*, *F. vasquezii*, and *F. yuvinkae* coexist in the same habitat. Apart from these areas, the distributional ranges of the different taxa show little overlap.

3.5 Character evolution

The states of ten selected morphological characters were coded (Tab. 3) and mapped onto the four-locus chloroplast DNA phylogeny, published in REX et al. (2009). Character-state transformations for six of these characters are summarized in Fig. 138.

Tab. 3: Encoded morphological character states within *Fosterella*.
  1 Habit: Caulescence: 0 acaulescent. 1 subcaulescent. 2 caulescent.
  2 Leaves: Margin: 0 entire. 1 serrate.
  3 Petals: Inflection during/after anthesis: 0 straight/straight. 1 slightly recurved/straight. 2 strongly recurved/straight. 3 recoiled like watchsprings/recoiled like watchsprings.
  4 Peduncle bracts: Margin: 0 entire. 1 serrate.
  5 Leaves: Arrangement: 0 flat rosette. 1 erect rosette. 2 leaves spirally arranged along elongated stem.
  6 Inflorescence: Density of indumentum: 0 glabrous/glabrescent. 1 sparsely lanate. 2 densely lanate.
  7 Leaves, abaxial: Indumentum: 0 tomentose by stellate trichomes. 1 scattered lepidote by peltate trichomes with dentate margin. 2 densely appressed lepidote by peltate trichomes with fimbriate margin. 3 covered by a thick layer of interwoven, peltate trichomes with fimbriate margin.
  8 Style: Stigma type: 0 simple-erect. 1 conduplicate-spiral.
  9 Inflorescence: Ramification: 0 raceme/compound raceme. 1 panicle.
  10 Petals: Colour: 0 white. 1 yellow. 2 red.

Species	1 Habit: Caulescence	2 Leaves: Margin	3 Petals: Inflection during/after anthesis	4 Peduncle bracts: Margin	5 Leaves: Arrangement	6 Inflorescence: Density of indumentum	7 Leaves, abaxial: Indumentum	8 Style: Stigma type	9 Inflorescence: Ramification	10 Petals: Colour
F. albicans	1	1	3	0	1	2	3	0	1	0
F. aletroides	0	0	3	0	1	1	1	0	0	0
F. batistana	0	0	1	0	0	1	1	0	0	0
F. caulescens	2	1	3	0	2	2	2	1	1	0
F. chaparensis	0	0	1	0	1	0	1	0	0	0
F. christophii	0	0	1	0	0	2	1	1	1	0
F. cotacajensis	2	1	3	0	2	0	3	0	1	0
F. elviragrossiae	0	0	1	0	0	0	2	0	0	0
F. floridensis	0	0	0	0	1	2	2	0	1	0
F. gracilis	0	0	1	0	0	0	0	1	1	1
F. graminea	0	1	3	0	1	0	2	0	1	0
F. hatschbachii	0	0	1	0	0	0	0	0	0	0
F. heterophylla	2	0	3	0	2	0	2	1	1	0
F. kroemeri	0	0	3	0	1	0	3	0	1	0
F. micrantha	0	0	1	0	0	1	1	1	1	0
F. nicoliana	0	0	1	0	1	1	1	0	0	0
F. pearcei	0	0	3	0	1	2	2	0	1	0
F. penduliflora	0	0	1	0	0	0	1	1	0	0
F. petiolata	0	0	3	0	1	1	2	0	1	0
F. rexiae	2	1	3	1	2	1	2	0	1	0
F. robertreadii	0	0	3	0	0	1	2	0	1	0
F. rojasii	0	0	3	0	0	0	3	0	1	0
F. rusbyi	0	1	3	0	0	0	3	0	1	0
F. schidosperma	0	0	3	0	1	0	1	0	1	0
F. spectabilis	0	0	0	0	0	0	1	0	0	2
F. vasquezii	0	0	3	0	0	0	3	0	1	0
F. villosula	0	0	1	0	0	2	1	1	1	0
F. weberbaueri	1	0	2	0	0	0	1	0	0	0
F. weddelliana	2	1	3	1	2	0	3	0	1	0
F. windischii	0	0	3	0	0	0	3	0	1	0
F. yuvinkae	0	0	1	0	0	0	0	1	1	0

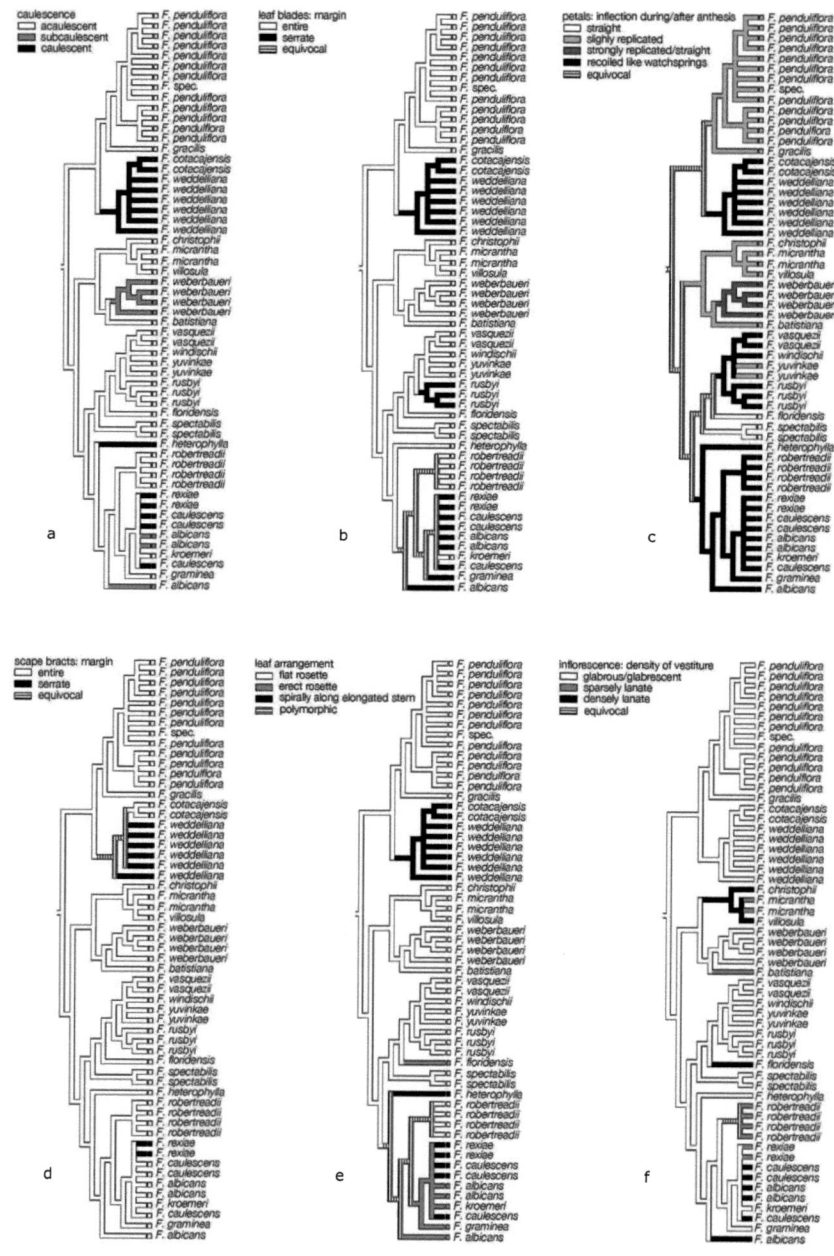

Fig. 138: Inference of character evolution in *Fosterella* using parsimony.
Morphological transitions for six selected characters (a–f) were mapped on the strict consensus tree of a parsimony ratchet analysis (adopted from REX et al. 2009).

4 Discussion

4.1 Collection efforts and taxonomical knowledge

Although *Fosterella* occurs in hardly explored countries with rather incomplete sampling coverage, it can be assumed that for Bolivia, the knowledge of the genus and its distribution is quite comprehensive. This is due to the fact that by now, *Fosterella* has been the focus of intensive and systematic research, also including collection of specimens, for more than ten years. Our interdisciplinary study of *Fosterella* taxonomy and phylogeny so far revealed the existence of morphological variability within taxa, but also indicated plausible species boundaries (IBISCH et al. 2002, 2006, 2008, submitted; REX et al. 2007, 2009; PETERS et al. 2008 a, b).

Both collection intensity and numbers of records per species are highest in the centres of species richness. Therefore, the probability of discovering additional, new *Fosterella* taxa in Bolivia is supposed to be rather low, with the possible exception of the largely unexplored and roadless Yungas forests. In the neighbouring countries of Bolivia, considerable less collection efforts have yet been dedicated to the genus *Fosterella*, and it is difficult to tell whether many new species will appear here. Presumably some species exclusively known from the La Paz department in Bolivia could be found in the Peruvian Yungas as well – or possibly even new species. The surprising record of *F. batistana* in the state of Pará in northern Brazil shows that this country, even besides the Brazilian shield, might be particularly promising for future collection efforts. Still another astonishing finding was *Fosterella nicoliana* from the Peruvian Amazon, which represents the northernmost record of the genus within the South American distribution range.

It is still an exciting question, whether the genus *Fosterella* is really absent from the ecoregions of the northern Andes. Theoretically, there should be more than enough adequate habitats, and obviously, one species (*F. micrantha*) even made it to Central America. To date, however, no *Fosterella* record is known from northern Peru, Ecuador, Colombia, Panama, Costa Rica, Nicaragua, Honduras or Belize. One cannot assume that all gaps regarding collection intensity can be closed in foreseeable future, but without doubt, the mentioned areas are auspicious for field observations, not only in respect of *Fosterella*.

4.2 Range sizes, ecoregions and ecology

The relatively small distribution ranges of most *Fosterella* species can be explained by their fragmented, island-like habitats like inter-Andean dry valleys (e.g., *F. cotacajensis*, *F. floridensis*) or Pre-Cambrian mountain ranges (*F. windischii*, *F. vasquezii*, *F. yuvinkae*). On the other hand, *Fosterella* species tend to have relatively large distributional ranges, when the habitats are continuous, like the Tucuman-Bolivian-Forest (*F. albicans*, *F. penduliflora*). Locally endemic species can be found in different ecoregions and at different altitudes.

The fact that the Yungas of the La Paz department, Bolivia, is the area with highest species richness is not very surprising, and can be explained by the extreme variable topography and extraordinary great habitat diversity of this region. Here, the river-shaped fragmentation of the broad eastern Andean slopes creates a small-scale mosaic of diverse habitats, ranging from humid-warm to dry-cool. The montane cloud forests of the tropical Andes are considered to be among the richest and most diverse forests on Earth. This makes sense in the light of increasing evidence that abiotic factors, like climate and topography, are good predictors of diversity patterns (KREFT & JETZ 2007).

4.3 Cytogenetics

Since there is now clear evidence for polyploidy in at least some *Fosterella* species, further investigation and a systematic survey are urgently needed to clarify the ploidy status in all *Fosterella* species, preferably by chromosome counts. Polyploidy in plants often is associated with high levels of genetic and morphological variation and evolutionary success (SOLTIS & SOLTIS 2000, ELLSTRAND & SCHIERENBECK 2000). Presently, there are several arguments in favour of the hypothesis that *F. penduliflora* is an allopolyploid hybrid that is currently undergoing diversification. These arguments come from (1) chromosome counts in the literature, (2) our own flow cytometry analyses and microsatellite data, (3) its widespread occurrence and morphological plasticity, and (4) the different position of the *penduliflora* group in the AFLP tree as compared with the cpDNA tree. Polyploidy might possibly be an important promoter of morphological and ecological plasticity and evolutionary success, in *F. penduliflora* and maybe in other *Fosterella* species as well. But hitherto the state of knowledge is too fragmentary to make definite statements.

4.4 Character evolution

Due to our fruitful cooperation dealing with the research subject *Fosterella*, including both molecular analyses and morphological studies, we are now in the fortunate position to compare the different results. The morphological data collected in the course of the present revision have been linked to the molecular findings of REX et al. (2009) by mapping selected character states onto a four-locus chloroplast phylogeny. This enabled us to draw conclusions about the ancestral versus derived state of several morphological characters. It also gave us an idea whether a given character state evolved only once, or several times independently. The different characters investigated here are discussed in detail below.

Caulescence

Most *Fosterella* species are acaulescent plants with a basal rosette. Distinct stems of more than 5 cm height have developed in five species only: *Fosterella caulescens*, *F. cotacajensis*, *F. heterophylla*, *F. rexiae* and *F. weddelliana*. In two species, *F. albicans* and *F. weberbaueri*, adult plants are subcaulescent, forming a rather inconspicuous stem of less than 5 cm height. The reconstruction of character evolution strongly suggests that acaulescence is the ancestral state within the genus (Fig. 87 a). Apparently, caulescence has evolved at least twice in the genus, once in the *weddelliana* group and once in the *albicans* group. Subcaulescence has been found in two other distinct lineages: in the *weberbaueri* group and within the *albicans* group, where it could be considered as an intermediate state.

Leaf blade margins

The majority of *Fosterella* species have entire leaf blades, but in seven species more or less serrate leaves can be found: *Fosterella albicans*, *F. caulescens*, *F. cotacajensis*, *F. graminea*, *F. rexiae*, *F. rusbyi* and *F. weddelliana*. Mapping this character onto the molecular phylogeny revealed that entire leaves are ancestral within the genus and serration is the derived state, which has evolved at least three times independently, i.e., within the *albicans*, the *rusbyi* and the *weddelliana* group (Fig. 87 b).

Inflection of petals

The late Robert W. Read considered the shape of the petals as a convenient character to divide the genus into two subgroups: one with petals recoiled like watchsprings during/after anthesis and the other with straight or lily-like petals becoming straight after anthesis (pers. comm. PIERRE L. IBISCH). Read even thought about a division into two subgenera based on this character, but never published his ideas. First molecular data resulting from AFLP-analysis seemed to support this idea (REX et al. 2007) but the chloroplast-based phylogeny challenged it (REX et al. 2009). Consequently, a different approach with a more detailed classification of the petal shape was taken here: Read's category "recoiled like watchsprings during anthesis and afterwards" was retained, but the category "straight or lily-like" was divided into (1) straight during anthesis and afterwards, (2) slightly replicated (lily-like) during anthesis and straight afterwards and (3) strongly replicated during anthesis and straight afterwards. Mapping the four character states onto the molecular phylogeny indicates that a principal subdivision of the genus based upon this character would not reflect natural groups (Fig. 87 c). Nonetheless, several groups are characterised by a common type of petal inflection: all species of the *weddelliana* and the *albicans* group have petals that are recoiled like watchsprings during anthesis and afterwards, whereas all species of the *penduliflora* and the *micrantha* group have slightly recurved petals during anthesis and afterwards. As the state at the base of the tree is equivocal, the question remains, which type of petal inflection might be regarded as ancestral. Apart from that, it becomes apparent that the different types of petal inflection have been gained or lost several times independently within the genus.

Peduncle bract margins

Almost all *Fosterella* species have entire peduncle bracts, except for *F. rexiae* and *F. weddelliana*, whose peduncle bracts are serrate. The inferred evolution implies that entire peduncle bracts are the ancestral state and serrate peduncle bracts evolved two times independently: once in the *albicans* group and once in the *weddelliana* group (Fig. 87 d).

Leaf arrangement

In the majority of *Fosterella* species, the leaves are arranged in a basal rosette. This can either be flat with leaves more or less adjacent to the ground or upright with more or less erect leaves not touching the ground. The leaves of caulescent species are spirally arranged along the stem. Character state reconstruction suggests that flat rosettes are ancestral within the genus (Fig. 87 e). Apparently, erect rosettes evolved at least twice, once in the *floridensis* group and once (maybe even twice) within the *albicans* group. Spiral leaf arrangement along with caulescent habit evolved within the *weddelliana* and the *albicans* group.

Indumentum of the inflorescence axes

The inflorescence axes of most *Fosterella* species are glabrous or glabrescent. Some species bear densely lanate inflorescences (*F. albicans, F. caulescens, christophii, F. floridensis, F. pearcei* and *F. villosula*) and some are only sparsely lanate (*F. aletroides, F. batistana, F. micrantha, F. nicoliana, F. petiolata, F. rexiae* and *F. robertreadii*). Character state reconstruction indicates that glabrous/glabrescent inflorescence axes are the ancestral condition within the genus and lanate inflorescences are derived (Fig. 87 f).

Mapping these character states onto the molecular phylogeny reveals that densely lanate inflorescences evolved several times independently: in the *micrantha* and *albicans* group as well as in *F. floridensis*. The same applies to sparsely lanate inflorescences, which are found in the *micrantha* group and in the *albicans* group as well as in *F. batistana*. A sparse indumentum can be considered as intermediate character.

Indumentum of the abaxial leaf surface

Almost all *Fosterella* species possess peltate trichomes of variable shape. Only three species are tomentose by stellate trichomes: *Fosterella gracilis*, *F. hatschbachii* and *F. yuvinkae*. The indumentum types with peltate trichomes were classified in (1) covered by a thick layer of interwoven, peltate trichomes with fimbriate margin, (2) densely appressed lepidote by peltate trichomes with fimbriate margin, and (3) scattered lepidote by peltate trichomes with dentate margin. Tracing the character onto the molecular phylogeny shows very high variability, and indicates that the latter type might be ancestral within the genus. Apparently, the other three indumentum types evolved several times independently but several species groups are uniform regarding this character. Within the *micrantha* and *weberbaueri* group all species are characterised by peltate trichomes with dentate margins, and the *weddelliana* group solely comprises species which are densely appressed lepidote by peltate trichomes with fimbriate margin.

Stigma type

The predominant stigma type within *Fosterella* is "simple-erect" (Type I according to BROWN & GILMARTIN 1984 b). Mapping the character on the phylogeny suggests that this is the ancestral character state within the genus. The "conduplicate-spiral" stigma type (Type II according to BROWN & GILMARTIN 1984 b) apparently is derived and evolved four times independently (once each within the *penduliflora*, *micrantha*, *rusbyi* and *albicans* group).

Ramification of the inflorescence

The majority of *Fosterella* species bear a panicle and only few species are characterised by a (compound) raceme. Inferred evolution implies that a panicle is the ancestral state within the genus and a (compound) raceme evolved independently in three different lineages, i.e., once each within the *penduliflora* and the *weberbaueri* group as well as in *F. spectabilis*.

Petal colour

All *Fosterella* species have white petals except for *F. gracilis* (yellow) and *F. spectabilis* (red). The molecular phylogeny indicates that white flowers are ancestral within the genus. In *F. spectabilis*, the red petal colour goes along with relatively long, tubular flowers and considerable nectar production. It is therefore assumed to be the only ornithophilous *Fosterella* species, whereas all other species presumably are entomophilous (KRÖMER 1997, IBISCH et al. 1999, IBISCH et al. 2002).

One special case, which unfortunately could not be included in the molecular studies because it is known from herbarium vouchers only, is represented by *Fosterella nicoliana*. This species stands out because of its globose fruits and clavate seeds which are very unusual for the genus. Hence it would be a rewarding project to visit the type locality, not only to confirm the record but also to get hold of living material of this remarkable species for molecular analyses.

4.5 Conclusions on the evolution and spread of the genus

Although a relatively well-resolved molecular phylogeny is now at hand for the genus *Fosterella* (REX et al. 2009), the tree topology does not tell us how the common ancestor of all *Fosterella* species might have looked like and whether it lived in humid or arid habitats. Nevertheless and in spite of its Andean centre of species richness, I assume that the genus has a lowland origin because of the following reasons: (1) The majority of *Fosterella* species show a mesomorphic habit and can be found in more or less humid habitats. (2) The genus is well-represented in the very old habitats of the Pre-Cambrian shield in the lowlands of central South America. (3) As far as we know, all *Fosterella* species carry out C3 photosynthesis, whereas the CAM (crassulacean acid metabolism) pathway is common in the highly xeromorphic genera of its sister group, comprising *Deuterocohnia*, *Dyckia* and *Encholirium*. In any case, the colonisation of Andean and/or lowland habitats most probably happened several times independently, maybe even in both directions.

The sister group relationship of *Fosterella* and the so-called *Dyckia* clade (comprising *Dyckia*, *Deuterocohnia* and *Encholirium*; CRAYN et al. 2004; REX et al. 2009) is also interesting regarding another aspect: given that sister groups by definition have the same age, the *Dyckia* clade with about 170 species has diversified much more than *Fosterella* with only about 30 species. This could be a result of a higher extinction rate in *Fosterella*, or of a higher speciation rate in the *Dyckia* clade, or both. One can only speculate about reasons for this disparity, but it is quite conceivable that the common ancestor of *Dyckia*, *Deuterocohnia* and *Encholirium* has gained the ability to perform CAM photosynthesis. This could have served as a key innovation enhancing diversification in discontinuous arid habitats, whereas *Fosterella* species – which lack this innovation – remained mostly associated with more or less humid habitats.

Fosterella micrantha, apparently a rather young taxon and closely related to *F. villosula*, is an obvious product of a relative recent event of long-distance dispersal (REX et al. 2007, 2009). This is interesting regarding two aspects: the habitats of *F. micrantha* in Central America are considerably drier than the humid montane forests, where *F. villosula* occurs. Beyond that, the distributional range of *F. micrantha* is remarkably large, as compared to that of the species of its sister group, *F. villosula* and *F. christophii*. Apparently, this taxon spread much more effectively after the initial founder event than most other *Fosterella* species in South America. Astonishingly, there is no evidence for a diversification of *F. micrantha* within Central America so far, although it arrived and became widespread in a new biogeographic region.

One could presume that the less rugged topography in conjunction with the ample availability of suitable habitats in northern Central America facilitates dispersal and gene flow as opposed to the topographically and ecologically diverse central South American Andes. If this argument holds, this would result in little genetic differentiation among *F. micrantha* populations, whereas differentiation between *F. villosula* and *F. christophii* is supposed to be relatively strong. The above hypothesis could be verified by phylogeographic analyses based on chloroplast DNA haplotypes in conjunction with phylogeographic analyses of AFLP patterns.

For the fact, that the majority of *Fosterella* species are characterised by small distribution ranges, there are two conceivable explanations: (a) The species are quite young, and (b) their dispersal mechanisms are quite ineffective. Given that the latter is the case, the distribution ranges of closely related species should be quite similar and close to each other. Although this seems to be true in some cases – *F. weddelliana* and *F. cotacajensis* as well as *F. caulescens* and *F. rexiae* coexist in ecologically similar habitats within the same geographical region – there are numerous exceptions. Thus, some closely related species pairs, like *F. vasquezii* and *F. rusbyi*, or *F. weberbaueri* and *F. batistana*, occur in geographically disjunct and ecologically distinct areas. Therefore, it can be assumed that long-distance-dispersal may happen, albeit quite seldom, and that narrow endemics do not necessarily have to be very young. However, the tiny *Fosterella* seeds definitely are unsuitable for efficient long-distance dispersal and such exceptional events seem to be rare

5 Summary

A comprehensive taxonomic revision of the neotropical genus *Fosterella* L.B. Sm. (Bromeliaceae) is presented, comprising all 31 currently accepted species and including a key to the species. The revision is based on traditional morphological examination of more than 800 herbarium specimens, about 150 living plants and own observations in the field.

The genus *Fosterella* has been the focus of an interdisciplinary study covering molecular as well as anatomical, morphological and biogeographical analyses. Our interest in *Fosterella* is fuelled by the ecological and biogeographical diversity of the genus. It is considered as an excellent model system for investigating speciation mechanisms in the Andes. In the last few years, various molecular methods have been applied to investigate interspecific relationships within the genus *Fosterella* so that by now well-resolved phylogenies are available. To complement these molecular studies which were mainly accomplished by Dr. Martina Rex, an intensive search of additional specimens was advanced and exhaustive taxonomical studies were initiated. In the course of these studies, the morphological variability of taxa and plausible species boundaries could be revealed, eventually resulting in a substantiated taxonomic concept of the genus.

Initially, a brief survey of the family Bromeliaceae as well as of the genus *Fosterella* is given, each including information on distribution, morphology, physiology, ecology and phylogeny. Subsequently, the taxonomic history of *Fosterella* is summarised, and the genus is delineated from the closely related genera *Deuterocohnia*, *Dyckia* and *Encholirium* which together form a sister group coined "*Dyckia* clade". The differential characters for delimitation of the individual *Fosterella* species are critically discussed with respect to their reliability and taxonomic relevance.

The key to the species of the genus *Fosterella* is based on easily detectable and scorable characters. For all species detailed descriptions are provided, including information on etymology, distribution, ecology, taxonomic delimitation and systematic relationships. All descriptions are accompanied by illustrations, a photo of the type, photos of living plants and a distribution map.

In the course of this revision, five species have been reduced to synonymy: thus, *Fosterella chiquitana* Ibisch, R. Vásquez & E. Gross and *F. latifolia* Ibisch, R. Vásquez & E. Gross have been reduced to synonymy of *F. penduliflora* (C.H. Wright) L.B. Sm.; *F. fuentesii* Ibisch, R. Vásquez & E. Gross to synonymy of *F. albicans* (Griseb.) L.B. Sm.; *F. elata* H. Luther to synonymy of *F. rusbyi* (Mez) L.B. Sm. and *F. nowickii* Ibisch, R. Vásquez & E. Gross to synonymy of *F. weddelliana* (Brongn. ex Baker) L.B. Sm.

Fosterella schidosperma (Baker) L.B. Sm. var. *vestita* L.B. Sm. & Read has been reduced to synonymy under *Fosterella weberbaueri* (Mez) L.B. Sm.

Six species have been described as new: *Fosterella batistana* Ibisch, Leme & J. Peters; *F. christophii* Ibisch, R. Vásquez & J. Peters; *F. elviragrossiae* Ibisch, R. Vásquez & J. Peters; *F. kroemeri* Ibisch, R. Vásquez & J. Peters; *F. nicoliana* J. Peters & Ibisch and *F. robertreadii* Ibisch & J. Peters. The taxon *F. gracilis* (Rusby) L.B. Sm. has been re-established.

In order to reconstruct character evolution, states of ten selected morphological characters were encoded and mapped on a molecular phylogeny. The following character states proved to be ancestral within the genus: aculescence, entire leaf blades, flat rosettes with leaves more or less adjacent to the ground, scattered lepidote abaxial leaf blades, peltate trichomes with a dentate margin, entire peduncle bracts, paniculate inflorescences, glabrous/glabrescent inflorescence axes, white petals and simple-erect stigmata.

Several conclusions regarding the evolution and spread of the genus could be drawn and were discussed in the context of the available literature: (1) The small distribution ranges that are characteristic for most *Fosterella* species apparently are linked to their fragmented, island-like suitable habitats, like inter-Andean dry valleys. (2) The fact that the Yungas of the La Paz department, Bolivia, is the region with the highest species richness can be explained by the extremely variable topography and the extraordinarily great habitat diversity of this region. (3) It seems to be most likely that the genus has a lowland origin because of the following reasons. The majority of *Fosterella* species have a rather mesomorphic habit and can be found in more or less humid habitats. The genus is well-represented in the very old habitats of the Pre-Cambrian shield in the lowlands of central South America. Furthermore, all *Fosterella* species carry out C3 photosynthesis as far as we know, whereas the CAM (crassulacean acid metabolism) pathway is common in the highly xeromorphic genera of the sister group, *Deuterocohnia*, *Dyckia* and *Encholirium*. Finally, the colonisation of Andean and/or lowland habitats by *Fosterella* species most probably happened several times independently, maybe even in both directions.

Zusammenfassung

Die vorliegende Arbeit stellt eine umfassende taxonomische Revision der Gattung *Fosterella* L.B. Sm. (Bromeliaceae) dar, die alle 31 derzeit akzeptierten Arten umfasst und einen Bestimmungsschlüssel für diese beinhaltet. Die Revision beruht auf der morphologisch-anatomischen Auswertung von Herbarmaterial (über 800 Exsikkate), Lebendpflanzen (ca. 150 Akzessionen) und eigenen vergleichenden Untersuchungen im Freiland.

Die Gattung *Fosterella* ist seit nunmehr etlichen Jahren Forschungsgegenstand einer interdisziplinären Studie, die sowohl molekulrae, als auch anatomische, morphologische und biogeographische

Untersuchungen einbezieht. Unser Interesse an der Gattung *Fosterella* gründet sich auf ihrer enormen ökologischen und biogeographischen Vielfalt, sie gilt als hervorragendes Modellsystem für Artbildungsmechanismen in den Anden. In den letzten Jahren wurde von verschiedenen molekularen Methoden Gebrauch gemacht, um die verwandtschaftlichen Beziehungen innerhalb der Gattung zu untersuchen, so dass mittlerweile gut aufgelöste Stammbäume vorliegen. Diese molekularen Studien, überwiegend durchgeführt von Dr. Martina Rex, wurden ergänzt durch intensive Sammelaktivitäten und eingehende taxonomische Untersuchungen im Rahmen der vorliegenden Revision. Auf diese Weise konnten die morphologische Plastizität der einzelnen Arten erfasst und schließlich ein wohlfundiertes Artkonzept vorgelegt werden.

Zunächst wird ein kurzer Überblick über die Familie der Bromeliaceen als auch die Gattung *Fosterella* gegeben, in dem jeweils Informationen zur Verbreitung, Morphologie, Physiologie, Ökologie und Phylogenie geliefert werden. Im Anschluss an einen historischen Überblick des taxonomischen Werdegangs wird die Abgrenzung der Gattung *Fosterella* zu den nächstverwandten Gattungen *Deuterocohnia*, *Dyckia* und *Encholirium* erläutert. Die morphologischen Merkmale zur Differenzierung der Arten innerhalb der Gattung werden im Hinblick auf ihre Zuverlässigkeit und ihr Gewicht diskutiert. Der Artschlüssel basiert auf Merkmalen, die leicht auszumachen und gut zu unterscheiden sind. Bei der ausführlichen Beschreibung der Arten wird auch auf ihre jeweilige Verbreitung, Ökologie, taxonomische Abgrenzung, systematische Verwandtschaft sowie die Etymologie des Namens eingegangen. Beigefügt sind jeweils Zeichnungen, ein Foto vom Holo-/Lectotypus, Fotos von Lebendpflanzen sowie eine Verbreitungskarte.

Im Rahmen der taxonomischen Arbeit wurden fünf Arten zu Synonymen reduziert: *Fosterella chiquitana* Ibisch, R. Vásquez & E. Gross und *F. latifolia* Ibisch, R. Vásquez & E. Gross wurden in die Synonymie von *F. penduliflora* (C.H. Wright) L.B. Sm. eingezogen; *F. fuentesii* Ibisch, R. Vásquez & E. Gross als Synonym zu *F. albicans* (Griseb.) L.B. Sm. gestellt; *F. elata* H. Luther in die Synonymie von *F. rusbyi* (Mez) L.B. Sm. verwiesen und *F. nowickii* Ibisch, R. Vásquez & E. Gross als Synonym zu *F. weddelliana* (Brongn. ex Baker) L.B. Sm. gestellt.

Fosterella schidosperma (Baker) L.B. Sm. var. *vestita* L.B. Sm. & Read wird zum Synonym von *Fosterella weberbaueri* (Mez) L.B. Sm. reduziert.

Sechs Arten wurden neu beschrieben: *Fosterella batistana* Ibisch, Leme & J. Peters; *F. christophii* Ibisch, R. Vásquez & J. Peters; *F. elviragrossiae* Ibisch, R. Vásquez & J. Peters; *F. kroemeri* Ibisch, R. Vásquez & J. Peters; *F. nicoliana* J. Peters & Ibisch und *F. robertreadii* Ibisch & J. Peters. Das Taxon *F. gracilis* (Rusby) L.B. Sm. wurde neu etabliert.

Um die Evolution von einzelnen morphologischen Merkmalen zu rekonstruieren, wurden die Zustände von zehn ausgewählten Merkmalen kodiert und auf einen molekularen Stammbaum kartiert. Die folgenden Merkmalszustände wurden als ursprünglich innerhalb der Gattung ermittelt: Stammlosigkeit, ganzrandige Blattspreiten, flache Rosetten mit dem Boden aufliegenden Blättern, locker beschuppte

Blattunterseiten, schildförmige Haare mit gezähntem Rand, ganzrandige Pedunkel-Brakteen, rispenförmiger Blütenstand, kahle/verkahlende Blütenstandsachsen, weiße Petalen und einfach-aufrechte Narben.

Rückschlüsse bezüglich der Evolution und Ausbreitung der Gattung *Fosterella* werden diskutiert: Die überwiegend kleinen Verbreitungsgebiete der Arten hängen offensichtlich mit ihren fragmentierten, inselartigen Habitaten (z.B. innerandine Trockentäler) zusammen. Die Tatsache, dass die Yungas-Bergregenwälder des Departamento La Paz, Bolivia, die Region mit der größten Artenvielfalt darstellen, lässt sich mit der extrem variablen Topographie und der außerordentlich hohen Vielfalt an Habitaten dieser Region erklären.

Aus folgenden Gründen erscheint es sehr wahrscheinlich, dass die Gattung *Fosterella* ihren Ursprung im Tiefland hat: Die Mehrheit der Arten weist einen eher mesophytischen Habitus auf und ist in mehr oder weniger humiden Habitaten zu finden. Die Gattung ist durch mehrere Arten in sehr alten Habitaten des präkambrischen Schilds im Tiefland von Zentral-Südamerika vertreten. Weiterhin betreiben, soweit bekannt, alle *Fosterella* Arten C3 Photosynthese, während in den Gattungen der Schwestergruppe, *Deuterocohnia*, *Dyckia* and *Encholirium*, CAM der verbreitete Photosyntheseweg ist. In jedem Fall ist die Besiedelung der Anden und/oder Tieflandhabitate mehrfach unabhängig voneinander geschehen, vielleicht sogar in beiden Richtungen.

6 References

APG – ANGIOSPERM PHYLOGENY GROUP (2003) An update of the Angiosperm Phylogeny Group classification for the orders and families of flowering plants: APG II. Botanical Journal of the Linnean Society 141: 399–436.

BAKER, J.G. (1876) Revision of the genera and species of Anthericeae and Eriospermae. Journal of the Linnean Society. Botany. 15: 326.

BAKER, J.G. (1889) Handbook of the Bromeliaceae. J. Cramer, 3301 Lehre.

BENZING, D.H. (1990) Vascular epiphytes. Cambridge University Press, New York.

BENZING, D.H. (2000) Bromeliaceae – Profile of an adaptive radiation. University Press, Cambridge: 98–105.

BÖHME, S. (1988) Bromelienstudien III. Vergleichende Untersuchungen zu Bau, Lage und systematischer Verwertbarkeit der Septalnektarien von Bromeliaceen. Tropische und subtropische Pflanzenwelt 62: 125–274.

BRAUER, I. (2006) Charakterisierung von Mikrosatelliten-DNA-Markern in der Gattung *Fosterella* (Bromeliaceae). Wissenschaftliche Hausarbeit zur Ersten Staatsprüfung für das Lehramt an Gymnasien, Universität Kassel.

BREMER, K. (2000) Early Cretaceous lineages of monocot flowering plants. Proceedings of the National Academy of Sciences of the United States of America 97: 4707–4711.

BROWN, G.K. & A.J. GILMARTIN (1984 a) Chromosome number reports LXXXV. Taxon 33 (4): 756–760.

BROWN, G.K. & A.J. GILMARTIN (1984 b) Stigma structure and variation in Bromeliaceae. Neglected taxonomic characters. Brittonia 36 (4): 364–374.

BROWN, G.K. & A.J. GILMARTIN (1986) Chromosomes of the Bromeliaceae. Selbyana 9: 88–93.

BROWN, G.K. & A.J. GILMARTIN (1989 a) Chromosome numbers in Bromeliaceae. American Journal of Botany 76 (5): 657–665.

BROWN, G.K. & A.J. GILMARTIN (1989 b) Stigma types in Bromeliaceae – a systematic survey. Systematic Botany 14 (1): 110–132.

BROWN, G.K. & R.G. TERRY (1992) Petal appendages in Bromeliaceae. American Journal of Botany 79 (9): 1051–1071.

BROWN, G.K., C.A. PALACÍ & H. LUTHER (1997) Chromosome numbers in Bromeliaceae. Selbyana 18 (1): 85–88.

BRUMMIT, R.K. & C.E. POWELL (1992) Authors of Plant Names. Royal Botanical Gardens, Kew.

BUDNOWSKI, A. (1922) Die Septaldrüsen der Bromeliaceen. Botanisches Archiv 1: 47–80, 101–105.

CHASE, M.W., D.E. SOLTIS, P.S. SOLTIS, P.J. RUDALL, M.F. FAY, W.H. HAHN, S. SULLIVAN, J. JOSEPH, M. MOLVRAY, P.J. KORES, T.J. GIVNISH, K.J. SYTSMA & J.C. PIRES (2000) Higher-level systematics of the Monocotyledons: an assessment of current knowledge and a new classification. In: WILSON K.L. & D.A. MORRISON (eds.) Monocots: Systematics and Evolution: 3–16. Proceedings of the 2nd International Monocot Symposium. CSRIO, Melbourn.

CHASE, M.W. (2004) Monocot relationships: an overview. American Journal of Botany 91 (10): 1645–1655.

CHASE, M.W., M.F. FAY, D.S. DEVEY, O. MAURIN, N. RØNSTED, T.J. DAVIES, Y. PILLON, G. PETERSEN, O. SEBERG, M.N. TAMURA, C.B. ASMUSSEN, K. HILU, T. BORSCH, J. DAVIS, D. STEVENSON, J.C. PIRES, T.J. GIVNISH, K.J. SYTSMA, M.A. MCPHERSON, S.W. GRAHAM & H. RAI (2006) Multigene analyses of monocot relationships: a summary. Aliso 22: 63–75.

CHAUDHRI, M.N., I.H. VEGTER & C.M. DEWAL (1972) Index herbariorum: A guide to the location and contents of the world's public herbaria. Part II (3). Collectors I–L.

CLARK, D.W., S.G. BRANDON, M.R. DUVALL, M.T. CLEGG (1993) Phylogenetic relationships of the Bromeliiflorae-Commeliniflorae-Zingiberiflorae-complex of monocots based on *rbc*L sequence comparisons. Annals of the Missouri Botanical Garden 80: 987–998.

COTIAS DE OLIVEIRA, A.L.P., J.G. DE A. ASSIS, M.C. BELLINTANI, J.C.S. ANDRADE & M.L.S. GUEDES (2000) Chromosome numbers in Bromeliaceae. Genetics and Molecular Biology 23 (1): 173–177.

COTIAS DE OLIVEIRA, A.L.P., J.G. DE A. ASSIS, G. DE O. CEITA, A.C.L. PALMEIRA & M.L.S. GUEDES (2004) Chromosome number for Bromeliaceae species occurring in Brazil. Cytologia 69 (2): 161–166.

CRAYN, D.M., K. WINTER & J.A. SMITH (2004) Multiple origins of crassulacean acid metabolism and the epiphytic habit in the Neotropical family Bromeliaceae. Proceedings of the National Academy of Sciences of the United States of America 101 (10): 3703–3708.

DAHLGREN, R.M.T., H.T. CLIFFORD & P.F. YEO (1985) The families of the Monocotyledons: Structure, evolution and taxonomy. Springer Verlag, Berlin.

DELAY (1974 a) Recherches sur la structure des noyaux quiescences chez les Phanérogames. Revue de Cytologie et de Cytophysiologie Vegetales 9: 169–222.

DELAY (1974 b) Recherches sur la structure des noyaux quiescences chez les Phanérogames. Revue de Cytologie et de Cytophysiologie Vegetales 10: 103–229.

DOUTRELIGNE (1939) Les diverse types de structure nucléare et de mitose somatique ches les Phanérogams. Cellule 2: 191–212.

DUVALL, M.R., M.T. CLEGG, M.W. CHASE, W.D. CLARK, W.J. KRESS, H.G. HILLS, L.E. EGUIARTE, J.F. SMITH, B.S. GAUT, E.A. ZIMMER & G.H. LEARN JR. (1993) Phylogenetic hypotheses for the monocotyledons constructed from *rbc*L sequence data. Annals of the Missouri Botanical Garden 80 (3): 607–619.

EHLER, N. & R. SCHILL (1973) Die Pollenmorphologie der Bromeliaceae. Pollen & Spores 15: 13–45.

EHLER, N. (1977) Bromelienstudien II: Neue Untersuchungen zur Entwicklung, Struktur und Funktion der Bromelien-Trichome. Tropische und Subtropische Pflanzenwelt 20: 473–508.

ELLSTRAND, N.C. & K.A. SCHIERENBECK (2000) Hybridization as a stimulus for the evolution of invasiveness in plants? Proceedings of the National Academy of Sciences of the United States of America 97: 7043–7050.

ESPEJO, A., A.R. LÓPEZ-FERRARI, I. RAMÍREZ-MORILLO, B.K. HOLST, H.E LUTHER & W. TILL. (2004) Checklist of Mexican Bromeliaceae with notes on species distribution and levels of endemism. Selbyana 25 (1): 33–86.

ESPEJO-SERNA, A., A.R. LÓPEZ-FERRARI & I. RAMÍREZ-MORILLO (2005) Bromeliaceae. Flora de Veracruz, Fasc. 136: 1–305.

ESPEJO-SERNA, A., A.R. LÓPEZ-FARRARI, N. MARTÍNEZ-CORREA & V.A. PULIDO-ESPARZA (2007) Bromeliad flora of Oaxaca, Mexico. Acta Botanica Mexicana 81: 71–147.

GILMARTIN A.J & G.K. BROWN (1987) Bromeliales, related monocots, and resolution of relationships among Bromeliaceae subfamilies. Systematic Botany 12: 493–500.

GITAÍ, J. R. HORRES & A.M. BENKO-ISEPPON (2005) Chromosomal features and evolution of Bromeliaceae. Plant Systematics and Evolution 253: 65–80.

GIVNISH, T.J., K.J. SYTSMA, J.F. SMITH, W.J. HAHN, D.H. BENZING & E.M. BURKHARDT (1997) Molecular evolution and adaptive radiation in *Brocchinia* (Bromeliaceae: Pitcairnioideae) atop tepuis of the Guayana Shield. In: GIVNISH, T.J. & K.J. SYTSMA (eds.) Molecular Evolution and Adaptive Radiation: 259–311. Cambridge University Press, Cambridge, UK.

GIVNISH, T.J., K.C. MILLAM, T.M. EVANS, J.C. HALL, J.C. PIRES, P.E. BERRY & K.J. SYTSMA (2004) Ancient vicariance or recent long-distance dispersal? Inferences about phylogeny and South American–African disjunctions in Rapateaceae and Bromeliaceae based on *ndh*F sequence data. International Journal of Plant Sciences 165 (4 Suppl.): S35–S54.

GIVNISH, T.J., J.C. PIRES, S.W. GRAHAM, M.A. MCPHERSON, L.M. PRINCE, T.B. PATTERSON, H.S. RAI, E.H. ROALSON, T.M. EVANS, W.J. HAHN, K.C. MILLAM, A.W. MEEROW, M. MOLVRAY, P.J. KORES, H.E. O'BRIEN, J.C. HALL, W.J. KRESS & K.J. SYTSMA (2006) Phylogenetic relationship of monocots based on the highly informative plastid gene ndhF: evidence for widespread concerted convergence. Aliso 22: 28–51.

GIVNISH, T.J., K.C. MILLAM, P.E. BERRY & K.J. SYTSMA (2007) Phylogeny, adaptive radiation, and historical biogeography of Bromeliaceae inferred from *ndh*F sequence data. Aliso 23: 3–26.

GRAHAM, S.W., J.M. ZGURSKI, M.A. MCPHERSON, D.M. CHERNIAWSKY, J.M. SAARELA, E.S.C. HORNE, S.Y. SMITH, W.A. WONG, H.E. O'BRIEN, V.L. BIRON, J.C. PIRES, R.G. OLMSTEAD, M.W. CHASE & H.S. RAI (2006) Robust inference of monocot deep phylogeny using an expanded multigene plastid data set. Aliso 22 (1): 3–20.

GRISEBACH, A.H.R. (1879) Symbolae ad Floram Argentinam. Abhandlungen der Königlichen Gesellschaft der Wissenschaften zu Göttingen, 24: 3–346.

GROSS, E. (1988) Bromelienstudien IV. Zur Morphologie der Bromeliaceen-Samen unter Berücksichtigung systematisch-taxonomischer Aspekte. Tropische und Subtropische Pflanzenwelt 64: 415–625.

GROSS, E. (1992) Die Samen der Bromeliaceae I. Die Bromelie 1992 (3): 61–66.

HALBRITTER, H. (1992) Morphologie und systematische Bedeutung des Pollens der Bromeliaceae. Grana 31: 197–212.

HARMS, H. (1929) Bromeliaceae novae III. Notizblatt 10. 1929: 794–795.

HARMS, H. (1930) Bromeliaceae. In: ENGLER, A. (ed) Die natürlichen Pflanzenfamilien. 2. Auflage. S. 65–158. Wilhelm Engelmann, Leipzig.

HOLMGREN, P.K., N.H. HOLMGREN & L.C. BARNETT (1990) Index Herbariorum, 1: The herbaria of the world. 8th ed. Regnum Vegetabile 120. New York Botanical Garden, New York.

HORRES, R. & G. ZIZKA (1995) Untersuchungen zur Blattsukkulent bei Bromeliaceae. Beiträge zur Biologie der Pflanzen 69: 43–76.

HORRES, R., G. ZIZKA, G. KAHL & K. WEISING (2000) Molecular phylogenetics of Bromeliaceae: evidence from *trn*L(UAA) intron sequences of the chloroplast genome. Plant Biology 2: 306–315.

HORRES, R., K. SCHULTE, K. WEISING & G. ZIZKA (2007) Systematics of Bromelioideae (Bromeliaceae) – evidence from molecular and anatomical studies. Aliso 23: 27–43.

IBISCH, P.L. (1996) Neotropische Epiphytendiversität – das Beispiel Bolivien. Martina-Galunder-Verlag, Wiehl.

IBISCH, P.L., E. GROSS, G. RAUER & D. RUDOLPH (1997). On the diversity and biogeography of the genus *Fosterella* L.B. Smith (Bromeliaceae). With the description of a new species from eastern Bolivia. Journal of the Bromeliad Society 47 (5): 211–217.

IBISCH, P.L., R. VÁSQUEZ & E. GROSS (1999) More novelties of *Fosterella* L.B. Smith (Bromeliaceae) from Bolivia. Revista de la Sociedad Boliviana Botánica 2 (2): 117–132.

IBISCH, P.L., C. NOWICKI, & R. VÁSQUEZ (2001) Towards an understanding of diversity patterns and conservation requirements of the Bolivian Bromeliaceae. Journal of the Bromeliad Society 51 (3): 99–113.

IBISCH, P.L., R. VÁSQUEZ, E. GROSS, T. KRÖMER & M. REX. (2002) Novelties in Bolivian *Fosterella* (Bromeliaceae). Selbyana 23 (2): 204–219.

IBISCH, P.L., S.G. BECK, B. GERKMANN & A. CARRETERO (2004 a) Ecoregions and ecosystems. In: IBISCH, P.L. & G. MÉRIDA (eds.) Biodiversity: The richness of Bolivia. State of Knowledge and Conservation: 47–88. Ministerio de Desarrollo Sostenible y Planificacion, Editorial F.A.N., Santa Cruz de la Sierra, Bolivia.

IBISCH, P.L., A. LEY, C. NOWICKI & R. VÁSQUEZ (2004 b) Diversity and biogeography of the Laeliinae, Polystachinae, Sobraliinae and Pleurothallidinae of Bolivia. In: VÁSQUEZ, R. & P.L. IBISCH (eds.) Orchids of Bolivia. Diversity and conservation status. Vol. 2. Laeliinae, Polystachyinae, Sobraliinae with update and complementation of the Pleurothallidinae: 39–68. Editorial F.A.N., Santa Cruz de la Sierra, Bolivia.

IBISCH, P.L., J. PETERS, M. REX, K. SCHULTE, A. OSINAGA & R. VÁSQUEZ (2006) Die Bromelien Boliviens (V): *Fosterella gracilis* (Rusby) L.B. Sm. Die Bromelie 2006 (2): 40–45.

IBISCH, P.I., J. PETERS, K. WEISING & R. VÁSQUEZ (in print) Die Bromelien Boliviens (VIII): Bemerkungen zu *Fosterella christophii, F. elviragrossiae, F. kroemeri, F. villosula und F. windischii*. Die Bromelie 2009 (2).

IBISCH, P.L., R. READ † & J. PETERS (2008) Key to the species of the genus *Fosterella* (Bromeliaceae). Selbyana 29 (2): 195–198.

KARL, D. (2008) Mikrostrukturelle Untersuchungen an Trichomen der Gattung *Fosterella* (Bromeliaceae). Wissenschaftliche Hausarbeit zur Ersten Staatsprüfung für das Lehramt an Gymnasien, Universität Kassel.

KESSLER, M., P.L. IBISCH & E. GROSS (1999) *Fosterella cotacajensis*, una nueva especie de Bromeliaceae de los valles secos andinos de Bolivia. Revista de la Sociedad Boliviana Botánica 2 (2): 111–116.

KESSLER, M. (2002 a) Environmental patterns and ecological correlates of range-size among bromeliad communities of Andean forests in Bolivia. The Botanical Review 68: 100–127.

KESSLER, M. (2002 b) Species richness and ecophysiological type among Bolivian bromelian communities. Biodiversity and Conservation 11: 987–1010.

KREFT, H. & W. JETZ (2007) Global patterns and determinants of vascular plant diversity. Proceedings of the National Academy of Sciences of the United States of America 104: 5925–5930.

KRÖMER, T. (1997) Untersuchungen zur Verbreitung, Ökologie und Nektarzusammensetzungvon Bromeliaceen. Diploma thesis, University of Göttingen.

KRÖMER, T., M. KESSLER, B.K. HOLST, H.E. LUTHER, E. GOUDA, P.L. IBISCH, W. TILL & R. VÁSQUEZ (1999) Checklist of Bolivian Bromeliaceae with notes on species distribution and levels of endemism. Selbyana 20 (2): 201–223.

LAWRENCE, G.H.M., A.F.G. BUCHHEIM, G.S. DANIELS & H. DOLEZAL (1968) Botanico Periodicum Huntianum. Hunt Botanical Library, Pittsburgh.

LECHLER, W. (1857) Berberides Americae australis. E. Schweizerbart, Stuttgart.

LINDLEY, J. (1843) Miscellaneous matter of the Botanical Register. Edwards's Botanical Register 29: 44.

LINDSCHAU, M. (1933) Beiträge zur Zytologie der Bromeliaceae. Planta 3: 506–530.

LUTHER, H. (1981) Miscellaneous new taxa of Bromeliaceae. Selbyana 5: 310–311.

LUTHER, H. (1979) Bromelienstudien I. Neue und wenig bekannte Arten aus Peru und anderen Ländern. Tropische und Subtropische Pflanzenwelt 31: 23–29.

LUTHER, H. (1997) A showy new *Fosterella* from Bolivia. Journal of the Bromeliad Society 47 (3): 118–119.

LUTHER, H. (2008) An alphabetical list of bromeliad binominals. Eleventh edition. Bromeliad Society International. Marie Selby Botanical Gardens, Sarasota, Florida.

MADDISON, D.R. & W.P. MADDISON (2003) MacClade 4: analysis of phylogeny and character evolution. Version 4.06. Sinauer Associates, Sunderland.

MARCHANT, C.J. (1967) Chromosome evolution in the Bromeliaceae. Kew Bulletin 21: 161–168.

MARTIN, C.E. (1994) Physiological Ecology of the Bromeliaceae. The Botanical Review 60 (1): 1–82.

MCNEILL, J., F.R. BARRIE, H.M. BURDET, V. DEMOULIN, D.L. HAWKSWORTH, K. MARHOLD, D.H. NICOLSON, J. PRADO, P.C. SILVA, J.E. SKOG, J.H. WIERSEMA & N.J. TURLAND (2006) International Code of Botanical Nomenclature (Vienna Code). Regnum Vegetabile 146. A.R.G. Gantner Verlag KG, Ruggell.

MCWILLIAMS, E. (1974) Chromosome number and evolution. In: SMITH L.B. & R.J. DOWNS (1974) Flora Neotropica. Monograph No. 14 (1): Pitcairnioideae (Bromeliaceae): 33–40. Haefner Press, New York.

MEZ, C. (1896) Bromeliaceae. In: DECANDOLLE, C. Monographiae Phanerogamarum 9. Sumptibus Masson & Co, Paris.

MEZ, C. (1901) Bromeliaceae et Lauraceae novae vel adhuc non satis cognitae. Botanische Jahrbücher für Systematik, Pflanzengeschichte und Pflanzengeographie 30. Beiblatt 67: 1–20.

MEZ, C. (1904) Additamenta monographica 1904. Bulletin de L'Herbier Boissier II. 4: 863–878.

MEZ, C. (1913) Additamenta monographica 1913. Repertorium specierum novarum regni vegetabilis 12: 417.

MEZ, C. (1934) Bromeliaceae. In: ENGLER, A. & L. DIELS (eds.) Das Pflanzenreich. IV. 32. Engelmann, Stuttgart. 1–667.

MORAWETZ, W. & C. RAEDIG (2007) Angiosperm biodiversity, endemism and conservation in the neotropics. Taxon 56 (4): 1245–1254.

MÜLLER, R., C. NOWICKI, W. BARTHLOTT & P.L. IBISCH (2003) Biodiversity and endemism mapping as a tool for regional conservation planning – Case study of the Pleurothallidinae (Orchidaceae) of the Andean rain forests in Bolivia. Biodiversity and Conservation 12 (10): 2005–2024.

NATIONAL GEOGRAPHIC SOCIETY & WWF (2009) WildWorld. Terrestrial ecoregions of the world (http://www.nationalgeographic.com/wildworld/terrestrial.html, last accessed May 26, 2009).

NOWICKI, C. (2004) Naturschutz in Raum und Zeit. Biodiversitätsextrapolationen, Klimaszenarien und soziodemographische Analysen als Instrumente der Naturschutzplanung am Beispiel Boliviens. GTZ. Eschborn.

PALMA-SILVA, C., D.G. DOS SANTOS, E. KALTCHUK-SANTOS & M.H. BODANESE-ZANETTI (2004) Chromosome numbers, meiotic behavior, and pollen viability of species of *Vriesea* and *Aechmea* genera (Bromeliaceae) native to Rio Grande do Sul, Brazil. American Journal of Botany 91 (6): 804–807.

PATZOLT, K. (2005) Blattanatomische Untersuchungen zur Systematic der Gattung *Fosterella* (Bromeliaceae). Diplomarbeit im Fachbereich Biologie und Informatik der Johann-Wolfgang-Goethe-Universität Frankfurt am Main.

PERCY NÚÑEZ, M., W. GALIANO, L. VALENZUELA, A. ZEGARRA, A. RODRÍGUEZ, E. SUCCLI & F. CARAZAS (2001) Estudio de Orchideas y otras plantas ornamentales promisorias de la Reserva de Biosfera y Parque Nacional del Manu. Informe No 002-2001. Edym Pro-Manu Principal.

PETERS, J. K. WEISING & P.L. IBISCH (2008 a) *Fosterella nicoliana*, eine neue Art des peruanischen Amazonas-Tieflands mit einem bislang unbekannten Samentyp, Die Bromelie 2008 (2): 64–69.

PETERS, J., R. VÁSQUEZ, A. OSINAGA, E. LEME, K. WEISING & P.L. IBISCH (2008 b) Towards a taxonomic revision of the genus *Fosterella* (Bromeliaceae). Selbyana 29 (2): 182–194.

PIERCE, S., K. MAXWELL, H. GRIFFITHS, K. WINTER (2001) Hydrophobic trichome layers and epicuticular wax powders in Bromeliaceae. American Journal of Botany 88: 1371–1388.

POREMBSKI, S. & W. BARTHLOTT (1999) *Pitcairnia feliciana* : The only indigenous African bromeliad. Harvard Papers of Botany 4 (1): 175–183.

RAFIQPOOR, D., C. NOWICKI, R. VILLARPANDO, A. JARVIS, H. SOMMER, P. JONES, P.L. IBISCH (2004) The factor that most influences the distribution of biodiversity: The climate. In: IBISCH, P.L. G. MÉRIDA (eds.) Biodiversity: The richness of Bolivia. State of Knowledge and Conservation. Ministerio de Desarrollo Sostenible y Planificacion / Editorial FAN, Santa Cruz.

RANKER, T.A., D.E. SOLTIS, P.S. SOLTIS & A.J. GILMARTIN (1990) Subfamilial phylogenetic relationship of the Bromeliaceae: evidence from chloroplast DNA restriction site variation. Systematic Botany 15 (3): 425–434.

RAUH, W. (1979) Bromelienstudien. Tropische und Subtropische Pflanzenwelt 31: 23–29.

RAUH, W. (1987) Bromelienstudien. Tropische und Subtropische Pflanzenwelt 60: 24–27.

REX, M., K. PATZOLT, K. SCHULTE, G. ZIZKA, R. VÁSQUEZ, P.L. IBISCH & K. WEISING (2007) AFLP analysis of genetic relationships in the genus *Fosterella* L.B. Smith (Pitcairnioideae, Bromeliaceae). Genome 50: 90–105.

REX, M., K. SCHULTE, G. ZIZKA, J. PETERS, R. VÁSQUEZ, P.L. IBISCH, K. WEISING (2009) Phylogenetic analysis of *Fosterella* L.B. Sm. (Pitcairnioideae, Bromeliaceae) based on four chloroplast DNA regions. Molecular Phylogenetics and Evolution 51: 472–485.

RUSBY, H.H. (1902) An enumeration of the plants collected by Dr. H.H. Rusby in South America, 1885–1886, XXXII. Bulletin of the Torrey Botanical Club 29: 697.

RUSBY, H.H. (1910) New species from Bolivia collected by R.S. Williams – I. Bulletin of the New York Botanical Garden 6: 487–517.

SCHILL, R., C. DANNENBAUM & E.-M. JENTSCH (1988) Untersuchungen an Bromeliennarben. Beitr. Biol. Pflanzen 63: 221–252.

SCHULTE, K, R. HORRES & G. ZIZKA (2005) Molecular phylogeny of Bromelioideae and its implications on biogeography and the evolution of CAM in the family (Poales, Bromeliaceae): Senckenbergiana Biologica 85: 113–125.

SIMPSON, M.G. (2006) Plant Systematics. Elsevier Academic Press, Burlington.

SMITH, L.B. (1934 a) Geographical evidence on the lines of evolution in the Bromeliaceae. Botanische Jahrbücher für Systematik, Pflanzengeschichte und Pflanzengeographie 66: 446–465.

SMITH, L.B. (1934 b) Studies in the Bromeliaceae – V. Contributions from the Gray Herbarium of Harvard University 104: 71–82.

SMITH, L.B. (1940) Notas sobre Bromeliáceas del Paraguay. Revista Argentina de Agronomía 7: 162–164.

SMITH, L.B. (1948) Bromeliaceaes notables de Bolivia. Lilloa 14: 93–96.

SMITH, L.B. (1954) Studies in the Bromeliaceae, XVII. Contributions from the United States National Herbarium 29: 521–543.

SMITH, L.B. (1960) Notes on Bromeliaceae XIV. Phytologia 7: 165–178.

Smith, L.B. (1962) Origins of the flora of Southern Brazil. Contributions from the United States National Herbarium 35 (3): 215–249.

Smith, L.B. (1963) Notes on Bromeliaceae, XIX. Phytologia 8: 407–510.

Smith, L.B. (1986) Revision of the Guayana Highland Bromeliaceae. Annals of the Missouri Botanical Garden: 659–721.

Smith, L.B. & R.J. Downs (1974) Flora Neotropica. Monograph No. 14 (1): Pitcairnioideae (Bromeliaceae). Haefner Press, New York.

Smith, L.B. & R.J. Downs (1977) Flora Neotropica. Monograph No. 14 (2): Tillandsioideae (Bromeliaceae). Haefner Press, New York.

Smith, L.B. & R.J. Downs (1979) Flora Neotropica. Monograph No. 14 (3): Bromelioideae (Bromeliaceae). Haefner Press, New York.

Smith, L.B. & R.W. Read (1982) Notes on Bromeliaceae, XLI. Phytologia 52 (1): 49–60.

Smith, L.B. & R.W. Read (1992) Flora Neotropica. Monograph No. 14 (1), Supplement No. 3. Bradea 15: 134–140.

Smith, L.B. & W. Till (1998) Bromeliaceae. In: Kubitzki, K. (ed.) The Families of Vascular Plants. Springer Verlag, Berlin: 74–99.

Soltis, P.S. & D.E. Soltis (2000) The role of genetic and genomic attributes in the success of polyploids. Proceedings of the National Academy of Sciences of the United States of America 97: 7051–7057.

Stafleu, F.A. & R.S. Cowan (1976–1988) Taxonomic literature 1–7. 2nd ed. Regnum Vegetabile. Bohn, Scheltema & Holkema, Utrecht.

Standley, P.C. (1923) New species of plants from Salvador. Journal of the Washington Academy of Sciences 13: 364.

Stapf, O. (1924) Plants from the Royal Botanic Gardens, Kew. Curtis's Botanical Magazine 150: pl. 9029.

Stearn, W.T. (2004) Botanical Latin. 4th ed. Timber Press, Portland.

Terry, R.G., G.K. Brown & R.G. Olmstead (1997) Examination of subfamilial phylogeny in Bromeliaceae using comparative sequencing of the plastid locus *ndh*F. American Journal of Botany 84 (5): 664–670.

Varadarajan, G.S. & G.K. Brown (1985) Chromosome number reports LXXXIX. Taxon 34 (4): 727–730.

Varadarajan, G.S. & A.J. Gilmartin (1987) Foliar scales of the subfamily Pitcairnioiseae (Bromeliaceae). Systematic Botany 12 (4): 562–571.

Varadarajan, G.S. & A.J. Gilmartin (1988 a) Seed morphology of the subfamily Pitcairnioideae (Bromeliaceae) and its systematic implications. American Journal of Botany 75 (6): 808–818.

VARADARAJAN, G.S. & A.J. GILMARTIN (1988 b) Phylogenetic relationships of groups of genera within the subfamily Pitcairnioideae (Bromeliaceae). Systematic Botany 13 (2): 283–293.

VARADARAJAN, G.S. & A.J. GILMARTIN (1988 c) Taxonomic realignments within the subfamily Pitcairnioideae (Bromeliaceae). Systematic Botany 13 (2): 294–299.

WRIGHT, C.H. (1910) Decades Kewensis: LVII. Bulletin of Miscellaneous Information, Royal Botanical Gardens, Kew: 192–197.

Die VDM Verlagsservicegesellschaft sucht für wissenschaftliche Verlage abgeschlossene und herausragende

Dissertationen, Habilitationen, Diplomarbeiten, Master Theses, Magisterarbeiten usw.

für die kostenlose Publikation als Fachbuch.

Sie verfügen über eine Arbeit, die hohen inhaltlichen und formalen Ansprüchen genügt, und haben Interesse an einer honorarvergüteten Publikation?

Dann senden Sie bitte erste Informationen über sich und Ihre Arbeit per Email an *info@vdm-vsg.de*.

Sie erhalten kurzfristig unser Feedback!

VDM Verlagsservicegesellschaft mbH
Dudweiler Landstr. 99　　　　　　　Telefon +49 681 3720 174
D - 66123 Saarbrücken　　　　　　 Fax　　 +49 681 3720 1749

www.vdm-vsg.de

Die VDM Verlagsservicegesellschaft mbH vertritt

Printed by Books on Demand GmbH, Norderstedt / Germany